Gas Turbine
Engineering Handbook

Gas Turbine Engineering Handbook

Editor

Sanjay Patil

Gas Turbine Engineering Handbook

Edited by **Sanjay Patil**

Printed in 2017

ISBN: 978-1-68117-384-9

Library of Congress Control Number: 2015936533

© 2016 by
SCITUS Academics LLC,
616, Corporate Way, Suite 2, 4766,
Valley Cottage, NY 10989

www.scitusacademics.com

Contents

Preface vii

Chapter 1   The Behaviour of Superalloys in Marine Gas Turbine Engine
             Conditions... 1
             I. Gurrappa, I. V. S. Yashwanth, and A. K. Gogia

Chapter 2   Testing of the Ultra-Micro Gas Turbine Devices (1 - 10 kW) for
             Portable Power Generation at University of Roma 1: First Tests
             Results... 17
             Alfonso Calabria, Roberto Capata, Mario Di Veroli, and
             Gianluca Pepe

Chapter 3   A Fault Diagnosis Approach for Gas Turbine Exhaust Gas
             Temperature Based on Fuzzy C-means Clustering and Support
             Vector Machine... 41
             Zhi-tao Wang, Ning-bo Zhao, Wei-ying Wang, Rui Tang, and
             Shu-ying Li

Chapter 4   Gas Turbine Blade Damper Optimization Methodology.................. 69
             R. K. Giridhar, P. V. Ramaiah, G. Krishnaiah, and S. G. Barad

Chapter 5   Fault Detection and Diagnosis for Gas Turbines Based on a
             Kernelized Information Entropy Model 109
             Weiying Wang, Zhiqiang Xu, Rui Tang, Shuying Li, and Wei Wu

Chapter 6   Surface Temperatures Determination with Influencing Convective
             and Radiative Thermal Resistance Parameters of Combustor of
             Gas Turbine 141
             Ebene Ufot, Ibiba Emmanuel Douglas, and Howel Iberefata Hart

Chapter 7   Dynamic Time-delay Characteristics and Structural Optimization
             Design of Marine Gas Turbine Intercooler 155
             Ning-bo Zhao, Xue-you Wen, and Shu-ying Li

vi

Chapter 8   **Gas Turbine Engine Control Design Using Fuzzy Logic and Neural Networks**..195

M. Bazazzadeh, H. Badihi, and A. Shahriari

Citations..225

Index..229

Preface

Gas turbine engineering handbook focuses on the design, fabrication, installation, operation, and maintenance of gas turbines. The third edition is not only an updating of the technology in gas turbines, which has seen a great leap forward in the 2000s, but also a rewriting of various sections to better answer today's problems in the design, fabrication, installation, operation, and maintenance of gas turbines. The third edition has added a new chapter that examines the case histories of gas turbines from deterioration of the performance of gas turbines to failures encountered in all the major components of the gas turbine.

Editor

Chapter 1

The Behaviour of Superalloys in Marine Gas Turbine Engine Conditions

I. Gurrappa[1], I. V. S. Yashwanth[2], and A. K. Gogia[1]

[1]Defence Metallurgical Research Laboratory, Kanchanbagh PO, Hyderabad, India
[2]M. V. S. R. Engineering College, Hyderabad, India

ABSTRACT

This paper presents hot corrosion results carried out systematically on the selected nickel based superalloys such as IN 738 LC, GTM-SU-718 and GTM-SU-263 for marine gas turbine engines both at high and low temperatures that represent type I and type II hot corrosion respectively. The results were compared with advanced superalloy under similar conditions in order to understand the characteristics of the selected superalloys. It is observed that the selected superalloys are relatively more resistant to type I and type II hot corrosion when compared to advanced superalloy. In fact, the advanced superalloy is extremely vulnerable to both types of hot corrosion. Subsequently, the relevant reaction mechanisms that are responsible for slow and faster degradation of various superalloys under varied hot corrosion conditions were discussed. Based on the results obtained with different techniques, a degradation mechanism for all the selected superalloys as well as advanced superalloy under both types of hot corrosion conditions was explained. Finally, the necessity as well as developmental efforts with regard to smart corrosion resistant coatings for their effective protection under high temperature conditions was stressed for their enhanced efficiency.

INTRODUCTION

Improved efficiency is the requirement for all types of modern gas turbines. In particular, achieving enhanced efficiency for marine gas turbines is a major challenge as the surrounding environment is highly aggressive. This aspect depends not only on the design but also on the selection of appropriate materials for their construction. Between the two, selection of materials plays a vital role as the materials have to perform well for the designed period under severe marine environmental conditions. Hot corrosion in a marine environment causes the materials to degrade at a significantly faster rate and causes catastrophic failures. It is important to mention that hot corrosion becomes a limiting factor for the life of components in marine gas turbines. Hence, the focus is on selection of appropriate materials and coatings.

Therefore, advanced materials with considerably improved properties are essential in order to enhance the efficiency of modern gas turbine engines. Efforts made in this direction made it possible to develop an advanced superalloy which exhibits excellent high temperature strength properties [1]. Application of high performance protective coatings over the conventional superalloys is an alternative approach to enhance the efficiency. As mentioned above, the issue is complicated for marine applications by the aggressivity of the environment [2-4]. Thus, the hot corrosion resistance of superalloys is as important as their high temperature strength in gas turbine engine applications [5-11]. An exhaustive review with recent developments as well as fundamentals of hot corrosion in gas turbine engines is available elsewhere [12].

Systematic studies were carried out on the selected superalloys in marine environments and at different elevated temperatures, which simulates the marine gas turbine engines. Comparative studies with an advanced superalloys were also carried out under similar environmental conditions in order to determine the nature of degradation and to establish the possible reaction mechanisms that cause the selected superalloys to corrode under marine environmental conditions before suggesting suitable high performance protective coatings for their effective protection.

EXPERIMENTAL

The selected superalloys for the present investigation are presented in Table 1. It is to be noted that the selected superalloys contain no rhenium but sufficient amount of chromium, while newly developed alloys contains about 6.5% rhenium and a very small amount of chromium. The modified chemistry with high contents of rhenium and tantalum makes the advanced superalloys to exhibit very good high temperature strength properties [1]. Small discs of about 3 mm thick were cut from all the superalloys, grounded up to 600 grit surface finish and cleaned with distilled water followed by acetone. Subsequently, the hot corrosion studies were carried out by a salt coating test, in which the specimens were coated with chloride and vanadium containing salts prior to hot corrosion studies at 900°C and 700°C, which represents type I and II hot corrosion respectively. Hot corrosion tests were also carried out at 800°C. The weight change data was recorded initially after 3 hours and later for every 20 hours. Each time, the specimens were washed with hot distilled water, dried and then recorded the weight. After recording the weight, the specimens were re-coated each time with the salt mixture and hot corrosion studies were carried out for a total period of 100 hours.

Table 1: The chemical composition of selected superalloys (wt %)

Superalloy	Ni	Cr	Co	W	Al	Ta	Ti	Mo	Re	Hf	Fe	Mn	Si	Cu
GTM-SU-263	Bal	20	20	-	0.6	1.3	2.4	6.0	-	-	0.7	0.6	0.4	0.2
GTM-SU-718	52.5	18.5	9.0	6.0	0.5	6.5	0.9	3.0	-	-	19.0	0.2	0.2	5.1 Nb
IN 738 LC	Bal	16	8.5	2.6	3.4	8.5	3.4	1.75	-	-	-	0.2	0.3	0.9 Nb
Advanced alloy	Bal	2.9	7.9	5.8	5.6	8.5	-	-	6.5	0.1	-			

After completion of hot corrosion tests, the specimens were examined for surface morphology with Scanning Electron Microscope (SEM) and the corrosion products were analyzed by Electron Dispersive Spectroscopy (EDS) and X-ray diffraction techniques. Cross sections of all the corroded specimens were analyzed for understanding the effect of hot corrosion and then elemental distribution was determined in order to evolve their degradation mechanisms.

RESULTS AND DISCUSSION

Figures 1, 2 and 3 show as hot corroded superalloys in chloride as well as vanadium containing environments under type I and type II hot corrosion conditions. As can be seen, all the selected superalloys were severely corroded at both the temperatures and environments. The corrosion is more severe under type I when compared to type II conditions in the both the environments. It indicates that all the superalloys are highly susceptible to hot corrosion. It is important to notice that the advanced superalloy is more vulnerable to hot corrosion (Figure 4) even under type II hot corrosion conditions i.e. at 700°C. Of course, the corrosion is much severe under type I Conditions i.e. 900°C. The advanced alloy degrades at a very faster rate making it difficult to recognize over a period of time as evidenced from the experiments. It is clearly indicating that the modified chemistry of the advanced superalloy could not improve its hot corrosion resistance. However, it exhibits very good high temperature strength characteristics. Whereas other selected superalloys exhibits relatively better hot corrosion resistance both under type I and type II conditions and their high temperature strength is less when compared to the advanced superalloy.

Figure 1: As hot corroded superalloy GTM-SU-263 in chloride and vanadium containing environments under type I and type II conditions.

Figure 5 demonstrates the weight change data for all the selected superalloys at 800°C for a period of 100 hours. The data revealed that all the superalloys were followed the same trend up to 20 hours and thereafter IN 738 LC superalloy forms the scale at a steady rate while other two superalloys form thicker scales and spalled subsequently. This behavior is same for every cycle and entire exposure period. The results obtained at other studied temperatures like 700°C and 900°C in both the environments exhibited similar behavior. It is clearly indicating that all the superalloys are not resistance to hot corrosion at all the temperatures in both the environments.

Figure 2: As hot corroded superalloy GTM-SU-718 in chloride and vanadium containing environments under type I and type II conditions.

Figure 3: As hot corroded superalloy IN 738 LC in chloride and vanadium containing environments under type I and type II conditions.

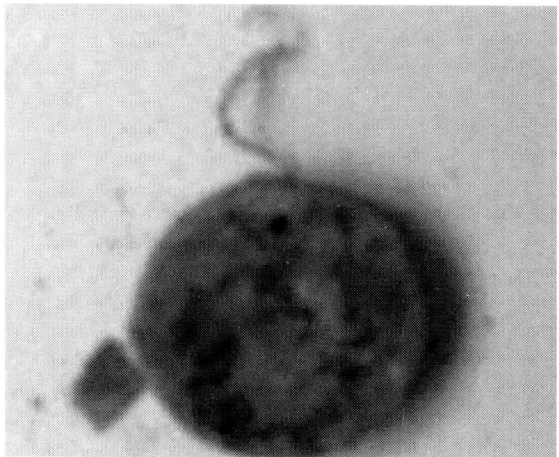

Figure 4: As hot corroded advanced super alloy in chloride containing environments under type II conditions.

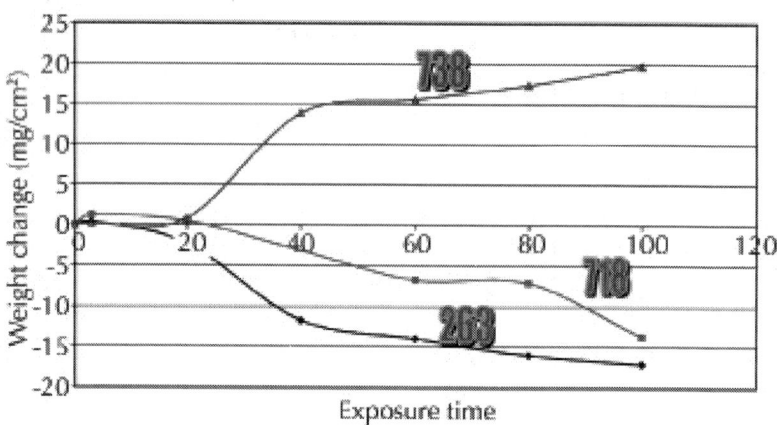

Figure 5: Weight change data for different superalloys at 800°C in vandium environment.

The surface morphologies of all the selected superalloys as well as advanced superalloy were studied and typical morphology of advanced superalloy at 900°C is presented in Figure 6. The surface morphology is different for different superalloys under the selected environmental conditions. ED's measurements revealed that the corrosion products contain sulphides and oxides of nickel and alloying elements of superalloys like Co, Cr, W, Ti, Ta, Re etc. The cross sections of hot corroded superalloys revealed that the corrosion-affected zone is large for all the superalloys. Among them, the affected zone is more for the advanced superalloy indicating that severe corrosion took place during the hot corrosion process for the advanced superalloy when compared to other selected superalloys (Figure 6).

The elemental distributions of all hot corroded superalloys including the advanced superalloy were studied in detail and typical elemental distribution of advanced superalloy at 900°C and 700°C are illustrated in Figures 7 and 8 respectively. In case of selected superalloys for marine gas turbines, which contains good amount of chromium could form continuous chromia scale on their surface. Particularly, for IN 738 LC, the continuous and protective chromia was clearly visible. They also promoted alumina as well as titania scales. However, diffusion of sulphur and oxygen into the superalloys was clearly observed. While SU 263 and 718, that contain less chromium could not form continuous chromia scale. Thin alumina scale was observed on the surface of superalloys. Small amounts of sodium and chlorine were also present in the corrosion products but not diffused into the superalloys. However, diffusion of sulphur and oxygen into the superalloys was noticed.

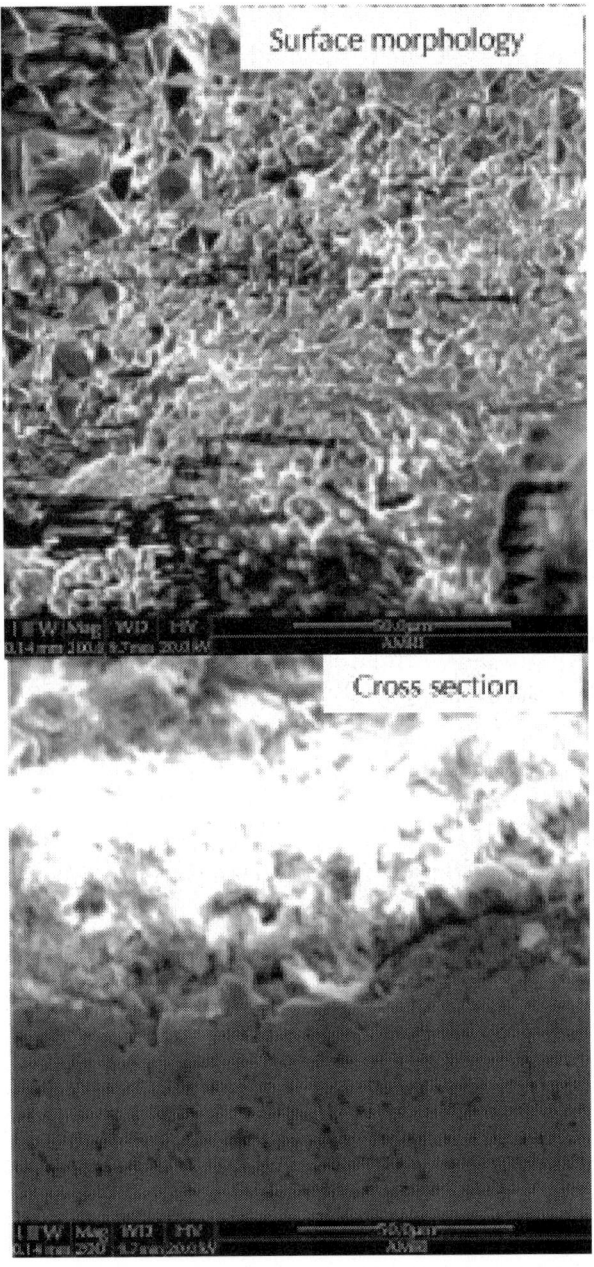

Figure 6: Surface morphology and cross section of advanced superalloy after type I hot corroison in chloride environment.

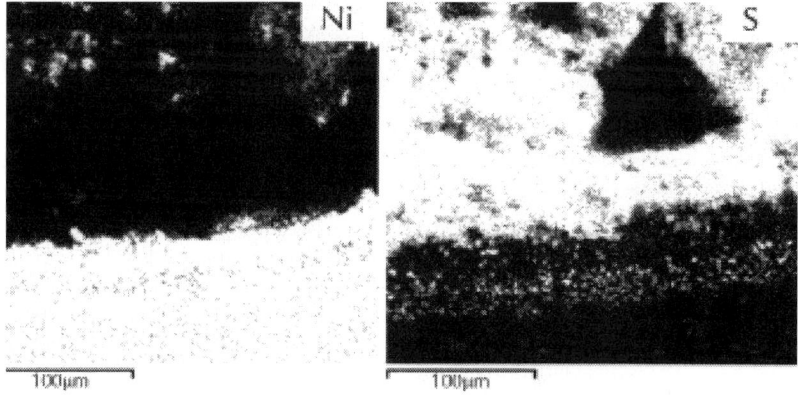

Figure 7: Elemental distribution of an advanced superalloy after hot corrosion at 900°C for 80 hours.

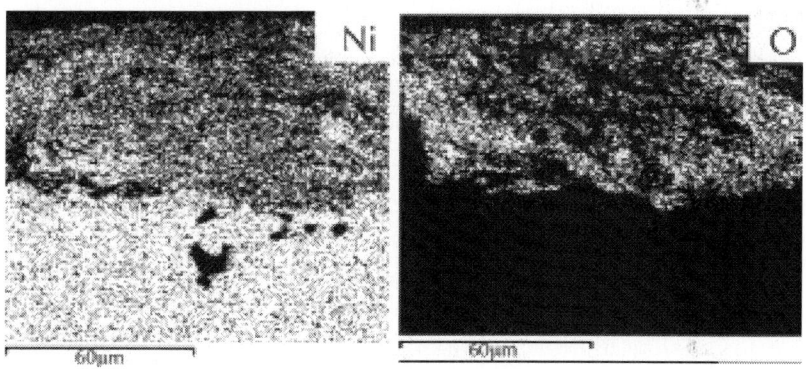

Figure 8: Elemental distribution of an advanced superalloy after hot corrosion at 700°C for 80 hours.

The elemental distribution of hot corroded advanced superalloy at 900°C and 700°C showed extensive presence of oxygen, sulphur and sodium in the corrosion products. Considerable diffusion of sulphur into the superalloy was clearly observed at 900°C (Figure 7) while oxygen at 700°C (Figure 8). Rhenium and tungsten were present in the corrosion products at 900°C and they were present in the corrosion affected zone of advanced superalloy that was hot corroded at 700°C. Ta and Hf were seen in the corrosion affected region. It is important to mention here that neither alumina nor chromia formation was observed on the superalloy.

It is due to the fact that chromium content in the adavanced superalloy is considerably low. At the same time, other alloying elements could not form any protective oxide scales. In essence, it is concluded that the advanced superalloy is highly susceptible to hot corrosion, though it exhibits excellent high temperature strength properties. It is important to mention here that the selected superalloys are also vulnerable to both types of hot corrosion but the intensity of attack is less. Among the selected superalloys, the IN 783 LC is more resistant while SU 263 is moderate and SU 718 is more susceptible. It clearly stresses the need to apply high performance protective coatings for their protection against hot corrosion both at low and high temperatures i.e. type II and type I as the marine gas turbine engines encounter both the problems during service. The protective coatings allow the marine gas turbine engines to operate at varied temperatures and enhance their efficiency by eliminating failures during service.

In essence, the present results clearly revealed that the selected superalloys as well as advanced superalloy are highly vulnerable to hot corrosion. The results further revealed that the advanced superalloy corrodes much faster when compared to selected superalloys. It is attributed to the fact that the tungsten which is the alloying element added along with other alloying elements in order to obtain high temperature strength characteristics of the superalloys, forms acidic tungsten oxide (WO_3) due to which fluxing of protective oxide scales such as alumina and chromia takes place very easily. This type of acidic fluxing is self-sustaining because WO_3 forms continuously that cause faster degradation of superalloys under marine environmental conditions at elevated temperatures. The degradation mechanism is explained in two steps as follows:

1. The tungsten present in the superalloys reacts with the oxide ions present in the environment and forms tungsten ion

$$WO_3 + O^{2-} = WO_4^{2-}$$

2. As a result, the oxide ion activity of the environment decreases to a level where acidic fluxing reaction with the protective alumina and chromia can occur

$$Al_2O_3 = Al^{3+} + O^{2-}$$

$$Cr_2O_3 = Cr^{3+} + O^{2-}$$

A similar reaction mechanism occurs if the superalloys contain other refractory elements like vanadium and molybdenum.

The following section describes an electrochemical phenomenon that explains the selected superalloys as well as advanced superalloy degradation process in detail under hot corrosion conditions:

Hot corrosion of all superalloys take place by oxidation of base as well as alloying elements like nickel, cobalt, chromium, aluminium, tantalum, rhenium etc. at the anodic site and forms Ni^{2+}, Co^{3+}, Cr^{3+}, Al^{3+}, Re^{4+}, Ta^{5+} ions etc. while at the cathodic site, SO_4^{2-} reduces to SO_3^{2-} or S or S^{2-} and oxygen to O^{2-}. Since the metal ions i.e. Ni^{2+}, Co^{3+}, Cr^{3+}, Al^{3+}, Re^{4+}, Ta^{5+} ions etc. are unstable at the elevated temperature and therefore reacts with the sulphur ions to form metal sulphides. The metal sulphides can easily undergo oxidation at elevated temperatures and form metal oxides by releasing free sulphur ($MS + 1/2O_2 = MO + S$). As a result, sulphur concentration increases at the surface of superalloys and enhances sulphur diffusion into them and form sulphides inside the superalloys. The practical observation of sulphides in hot corroded superalloys specimens clearly indicates that the electrochemical reactions took place during their hot corrosion process. Simultaneously, the metal ions react with oxide ions that are evolved at the cathodic site leading to the formation of metal oxides. The metal oxides dissociate at elevated temperatures to form metal ions and oxide ions. As a result, oxygen concentration increases at the surface and thereby diffuses into the superalloys. Practical observation of oxides in hot corroded superalloys specimens is a clear indication that the electrochemical reactions took place during the hot corrosion process.

Therefore, the hot corrosion of all the alloys i.e. selected and advanced superalloy, is electrochemical in nature and the relevant electrochemical reactions are shown below:

At the anode	At the cathode
$Ni = Ni^{2+} + 2e^-$	$1/2O_2 + 2e^- = O^{2-}$
$Cr = Cr^{3+} + 3e^-$	$SO_4^{2-} + 2e^- = SO_3^{2-} + O^{2-}$
$Co = Co^{3+} + 3e^-$	$SO_4^{2-} + 6e^- = S + 4O^{2-}$
$Al = Al^{3+} + 3e^-$	$SO_4^{2-} + 8e^- = S^{2-} + 4O^{2-}$
$Re = Re^{4+} + 4e^-$	
$Ta = Ta^{5+} + 5e^-$	

Figure 9 illustrates an electrochemical model showing that all the studied superalloys degradation is electrochemical in nature. Similar mechanism is applicable to other superalloys and their families.

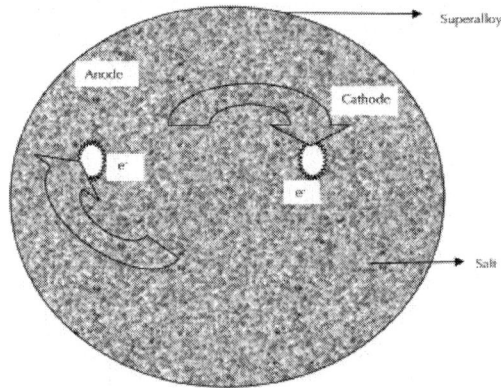

Figure 9: An electrochemical model showing the hot corrosion of selected and advanced superalloys is an electrochemical process.

The template is used to format your paper and style the text. All margins, column widths, line spaces, and text fonts are prescribed; please do not alter them. You may note peculiarities. For example, the head margin in this template measures proportionately more than is customary. This measurement and others are deliberate, using specifications that anticipate your paper as one part of the entire proceedings, and not as an independent document. Please do not revise any of the current designations.

DESIGN AND DEVELOPMENT OF SMART COATINGS

Recent extensive research has resulted in design and development of smart coatings which provide effective protection to the superalloy blades for the designed period against type I and type II hot corrosion that are normally encountered in marine and industrial gas turbine engines which in turn enhances their efficiency considerably [13,14]. The same coating can be applied to aero gas turbine engine components due the fact that hot corrosion is a concern when they move at low altitudes across the sea and provide total protection. This is a major developmental work in the area of gas turbine engines used in aero, marine and industrial applications. Unlike the conventional/existing coatings, the smart coatings provide total protection to the superalloy components used in aero, marine and industrial applications by forming appropriate protective scales depending on the surrounding environmental conditions [12-14].

REFERENCES

1. N. Das, US patent 5,925,198, July 1999

2. M. R. Khajavi and M. H. Shariat, "Failure of First Stage Gas Turbine Blades," Engineering Failure Analysis, Vol. 11, No. 4, 2004, pp. 589-597.doi:10.1016/j.engfailanal.2003.08.010

3. J. M. Gallardo, J. A. Rodrigue and E. J. Herrera, "Failure of Gas Turbine Blades," Wear, Vol. 252, No. 3-4, 2002, pp. 264-268. doi:10.1016/S0043-1648(01)00885-7

4.   N. Eliaz, G, Shemesh and R. M. Latarision, "Hot Corrosion in Gas Turbine Components," Engineering Failure Analysis, Vol. 9, No. 1, 2002, pp. 31-43. doi:10.1016/S1350-6307(00)00035-2

5.   M. Konter and M. Thumann, "Materials and Manufacturing of Advanced Industrial Gas Turbine Components," Journal of Materials Processing Technology, Vol. 117, No. 3, 2001, pp. 386-390. doi:10.1016/S0924-0136(01)00785-3

6.   J. Stringer, "High Temperature Corrosion of Superalloys," Materials Science and Technology, Vol. 3, 1987, pp. 482- 493

7.   A. S. Radcliff, "Factors Influencing Gas Turbine Use and Performance," Materials Science and Technology, Vol. 3, 1987, pp. 554-561.

8.   R. F. Singer, "New Materials for Industrial Gas Tubines," Materials Science and Technology, Vol. 3, 1987, pp. 726- 732.

9.   I. Gurrappa, "Hot Corrosion Behaviour of CM 247 LC Alloy in Na_2SO_4 and NaCl Environments," Oxidation of Metals, Vol. 51, No. 5-6, 1999, pp. 353-382.doi:10.1023/A:1018831025272

10.   C. J. Wang and J. H. Lin, "The Oxidation of MAR M247 Superalloy with Na_2SO_4 Coating," Materials Chemistry and Physics, Vol. 76, No. 2, 2002, pp. 123-129. doi:10.1016/S0254-0584(01)00527-2

11.   I. Gurrappa and A. S. Rao, "Thermal Barrier Coatings for Enhanced Efficiency of Gas Turbine Engines," Surface and Coating Technology, Vol. 201, No. 6, 2006, pp. 3016- 3029. doi:10.1016/j.surfcoat.2006.06.026

12.   I. Gurrappa, I. V. S. Yashwanth, A. K. Gogia, H. Murakami and S. Kuroda, Interlational Materials Revews. (In Press)

13.   I. Gurrappa, "Identification of a Smart Bond Coating for Gas Turbine Engine Applied Lications," Journal of Coating Technology Research, Vol. 5, No. 3, 2008, pp. 385- 390.doi:10.1007/s11998-008-9103-y

14.   I. Gurrappa, "Final Report on Design and Development of Smart Coatings for Aerospace Applied Lications," European Commission, July 2008

Testing of the Ultra-Micro Gas Turbine Devices (1 - 10 kW) for Portable Power Generation at University of Roma 1: First Tests Results

Alfonso Calabria, Roberto Capata, Mario Di Veroli,
and Gianluca Pepe

Department of Mechanical and Aerospace Engineering, Sapienza—University of Roma, Roma, Italy

ABSTRACT

The ever increasing development of portable electronics has led to a higher demand for compact and reliable power sources. Significant resources are being presently dedicated to the study of micro machined gas turbines, because of their remarkable power density. The paper reports the procedures and the results of a series of tests conducted at the Department of Mechanical & Aerospace Engineering of University of Roma 1, to obtain the map of an ultra-micro gas turbine device, and the head losses and the combustion efficiency of the corresponding ultra-micro combustion chamber, fed by a mixture of butane and propane.

This work is a part of a research aimed at the conception, design and prototyping of an ultra-micro thermo-electrical device for portable power generation. The novelty of the research consists in the fact that the thermal engine is a (ultra-micro) gas turbine set. In a subsequent stage, several different configurations have been assessed to select the most proper geometry and structural characteristics of the most relevant components (radial compressor, radial turbine, combustion chamber, electric motor and generator, bearings, regenerative heat exchanger).

INTRODUCTION

Although the quest for ultra-micro (and even nano-micro) portable power supply systems is over a decade old, and since the mid 90's several research teams all around the world have been actively involved in this very specialist field, today there are only a few and incomplete prototypes of such devices, most of them failing to reach acceptable performance. The fundamental idea is that of powering a micro-scale electric generator with a thermal device that can run on a generally available fuel (kerosene, propane/butane mixtures, methane, hydrogen, etc.), so that the user can avoid carrying a quite heavy battery package. Reciprocating engines being unfeasible at these small scales, the use of gas turbine micro-plants has been investigated. Major advances have been made in the US (MIT, Stanford [1,2]), in Japan [3] in Belgium (Leuven [4]) and in Italy, University of Roma 1 [5,6] and the work at ONERA laboratories in France.

All teams have until now adopted the same configuration as the one used in "large scale" gas turbines. There are other configurationally variations of this cycle, but for very small units, the simple Braytoncycle is the only one considered to date. For thermo-fluid dynamic reasons, the efficiency of a gas turbine set can be increased by either one of the following actions:

- Increase the maximum allowable combustion gas temperature, TIT;
- Correspondingly increase the pressure ratio of the compressor C;
- Optimize the regeneration (i.e., the amount of heat subtracted to the hot gases and delivered to the cold compressed air before CC) in the heat exchanger R.

The increase of the TIT (turbine inlet temperature) is limited solely by technological reasons: the blades of the turbine are usually made of special alloys that can sustain continued operation at a TIT of about 1300 K. For higher gas temperatures, internally cooled blades have been employed, or alternatively ceramic materials have been used to manufacture the blades. In ultra-small GT, blade cooling is technically impossible, while ceramics have demonstrated to be unsuitable for the extremely high rotational speeds required by these devices. The pressure ratio of the radial compressor is basically dictated by its rotational speed U, to whose square the specific work of the compressor (and hence the pressure ratio) is proportional. Thus, a higher β require higher U, and this poses both fluid dynamic (losses, leakage, efficiency) and structural (creep, vibrations) problems of difficult solution.

PRELIMINARY CONSIDERATIONS ON THE TESTS CAMPAIGN

The Experimental texts reported in this paper are a direct consequence of two projects being developed at the Department of Mechanical and Aerospace Engineering of Roma 1. In the first case, evaluation tests on the JETCAT model fall in "SEALAB" project for the construction of a high-performance marine vehicle. In the second case, the thermal mapping of combustion chamber and its specific consumption, within the project PNR 714 with Italian Army Head Quarter for the realization of a portable device. Both devices, then, could be used, once a possible connection with inverter has been realized, as a rangeextender on hybrid vehicles, UAVs or boats.

THE JETCAT GAS TURBINE (GT) DEVICE

The Graupner/Jetcat turbo-prop engine represents a successful combination of high power reserves and hightech engineering. In the world of full-size aviation most types of propeller-driven machine have already been converted to turbo-prop power, but the engine relentless progresses has only just begun in the model aviation arena. As the name indicates, the turbo-prop engine—its full name is a turbo-jet propeller engine—comprises a gas turbine driving an airscrew. The primary advantages of the turbo-prop in full-size aviation lie in its compact shape and its economy and reliability at speeds below 700 km/hr. The basic design and method of working of the new model engine correspond quite closely to those of the full-size power plants. The principle is very easy to understand: it is simply a matter of finding a suitable method of converting the high power of the turbine engine into usable shaft power. However, this is not necessarily a straightforward matter, especially when you consider the very high turbine rotational speed. In the engine presented here the rotational speed is reduced in two stages: the first is a gas reduction, the second a geared reduction. This means that when the turbine is running, the gas flow from the core engine drives a turbine wheel which is mounted on a second shaft. This second shaft is mechanically completely independent of the rotor of the base engine and not connected to it; it receives its power solely from the kinetic energy of the exhaust gas flow. The secondary shaft directly drives a gearbox designed to cope with high rotational speeds, and this in turn reduces the speed to a value suitable for a propeller. The gearbox is fitted with an integral axial fan which provides the necessary airflow for cooling the components exposed at high temperature. Another completely new feature of the engine is the electronic control system, which processes the speed information derived from both shafts, primary and secondary. This simply means that the pilot can concentrate entirely on flying, while the complex engine management processes are carried out fully electronically. The reduction gearbox is a specially developed planetary gear design, highly efficient and very compact. Whenever a new kind of power system concept is introduced, the model flyer is obliged to immerse himself in the subject in order to gain the necessary expertise, and this certainly applies to turbines. However, once the operator has become familiar with the procedure, it is actually simpler to handle a turbine installed in a model aircraft than to operate a piston engine: only a single radio control system channel is required to control the engine, and starting preparations for the engine simply boil down to filling the fuel tank and a small auxiliary gas tank required to start the turbine. The engine starts at the press of a button from the transmitter, and the entire starting process runs automatically, controlled by the gas turbine on-board electronics (ECU). Initially the integral electric motor accelerates the turbine to a speed of around 6000 rpm, then the auxiliary gas supply is opened and the gas is ignited in the turbine's combustor. The gas turbine then continues to accelerate until the burning gas overtakes the starter motor's speed; the motor then disengages, and the turbine continues to accelerate until it reaches a speed high enough to support running on kerosene. Once the start-up process is completed successfully, the ECU sets a stable idle speed before transferring control to the pilot. After the flight the pilot reduces engine speed in the usual way, then—using the same channel—he initiates the power-down process, which is again entirely automatic, under the control of the on-board electronics: first combustion is halted, then the starter motor is switched on again to push fresh air through the turbine until the internal temperature has fallen down to below 100°C. An LED in the case of model lights up to indicate that this cooling-off phase is complete, at which point the receiving system can safely be switched off.

The characteristics of the GT device are reported in the Table 1, while in Figure 1 is shown the device.

Jetcat Fuel Supply System

The Jetcat engine can use deodorized kerosene, 1-K kerosene or Jet-A1 as fuel. Fuel must be mixed with 5% synthetic turbine oil. Jetcat itself recommends Aeroshell 500 turbine oil although any turbine oil that conforms to

Figure 1: Jetcat model and accessories.

Table 1: GT published nameplate

Technical Characteristics	
Dimensions	330 x 130 x 275 mm
Rpm	30.000 - 112.000
Fuel consumption	130 - 700 mlimin
Fuel	Jet Al, Kerosene. Petroleum
Maintenance	every 50 hours
Exhaust temperature	every 50 hours
Weight	2359 g

MS23699 standards will work. The input and output fuel piping must be connected to the electronic shut-off valve as shows in Figures 2 and 3.

The tube from the pump, called "fuel in", is direct to the heat shrink pipe covered coil. The tube to the engine, "fuel out", is close to the edge of the valve. The engine is equipped with a smaller valve (brass/plastic body). The manufacturer recommends checking and clean the fuel filters every ten (10) flights/runs. When the engine runs at full power, the control unit ECU checks the fuel line from the pump to the engine. If there is a large quantity of air bubbles flowing with the fuel, probably due to a restriction in the fuel system or an air leak in a fitting, the system automatic shutdown. It is important not to over-pressurize the kerosene tanks and the kerosene shut off valve during refuelling operations. To prime the fuel pump and fuel lines (or for fuel pump test purposes), it is necessary to open the fuel shutoff valve and run the fuel pump manually.

Pump/Purge fuel operation allows the fuel pump to operate without the turbine running. However, if the fuel feed line is not removed from the turbine during this procedure; it will become flooded by fuel. When this occurs,

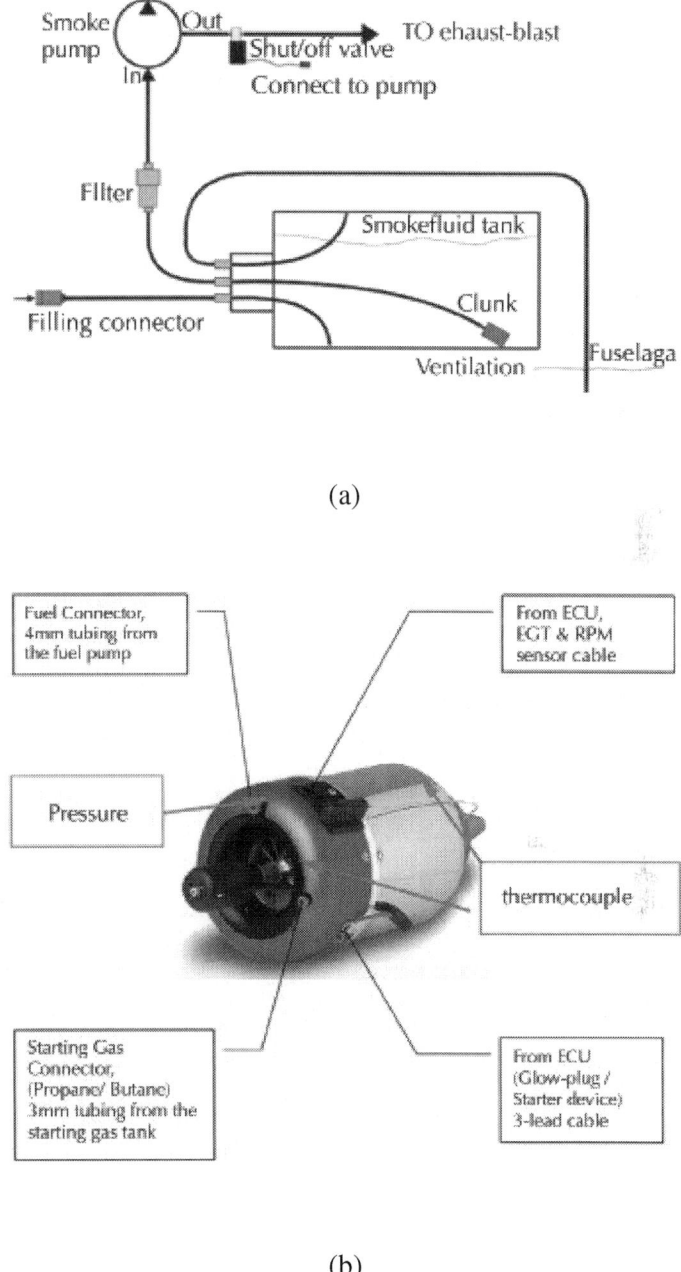

(a)

(b)

Figure 2: Jetcat fuel system and Jetcat connections points.

the next turbine start will be characterized by a higher fuel consumption.

Figure 2 also shows the additional measurement points used to map the compressor. The housing for the additional instruments has been created ad hoc, on the case of the distributor and the return channel.

The inlet conditions were considered those of the environment. In this case, the measure of temperature and ambient pressure is repeated many times using commercial tools not overly sophisticated. However, the value of the ambient temperature has fluctuated between 12 and 17°C and the pressure between 100.67 and 102.1 kPa. Regarding the operating speed and fuel consumption, this derives from the power control software using turbine ECU.

The devices used were a pressure sensor and two thermocouples Thanks to these measures we could compute the compressor pressure ratio and calculate, once known the discharge temperature, the efficiency for various speeds. These results are reported in Figures 4-7.

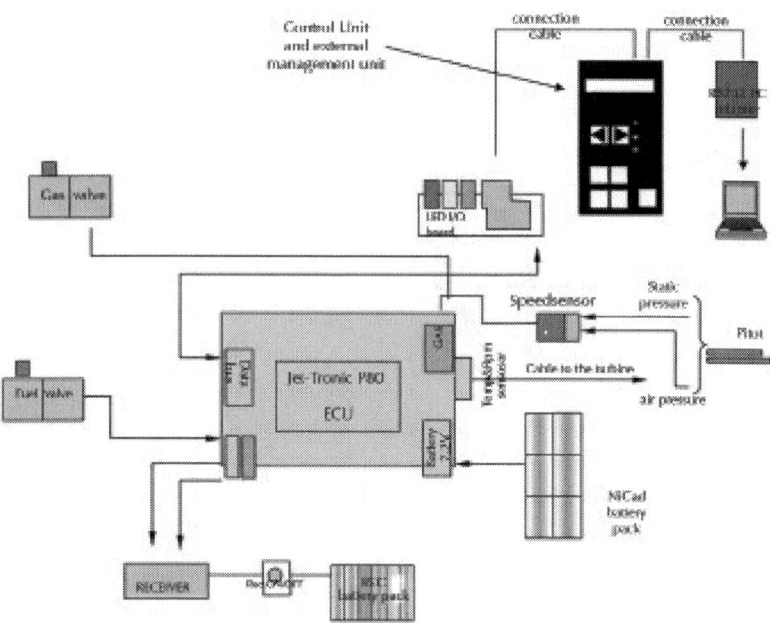

Figure 3: Jetcat connection arrangement and measurements standard devices supplied devices.

After the turbine has ignited, the starter motor further accelerates the turbine. At approximately 5000 RPM, the fuel pump is automatically started at its minimum power by the ECU. Beginning from this first pump start voltage, the fuel flow is then slowly increased by increasing the pump voltage.

The initial pump voltage that is supplied for the pump immediately after ignition has been adjusted by the factory. If the fuel pump is changed or after several turbine runs the pump is breaking in thus delivering too much fuel at start-up causing long flames behind the turbine exhaust, it might be necessary to readjust the pump start voltage. Table 2 reports the characteristics of the tests carried on the Jetcat GT device.

At the end of the tests it can establish that the compressor maximum pressure ratio is approximately equal to 3 for a mass flow rate of 0.30 kg/s. The efficiency of the compressor varies between 0.52 and 0.77.

Consequently the fuel consumption varies within 0.0015 to 0.010 kg/s. The turbine inlet temperature (TIT) varies within 858 to 1137 K.

THE ULTRA MICRO GAS TURBINE GENERATOR (UMGTG) AT UNIVERSITY OF ROMA 1 (UDR1)

The UMGTG-UDR1 device consists of the following fundamental components: compressor C, turbine T, com-

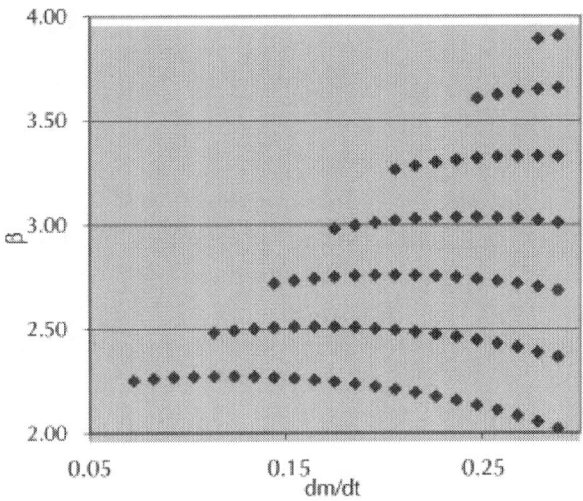

Figure 4: Compressor pressure ratio in function of compressor inlet mass flow rate.

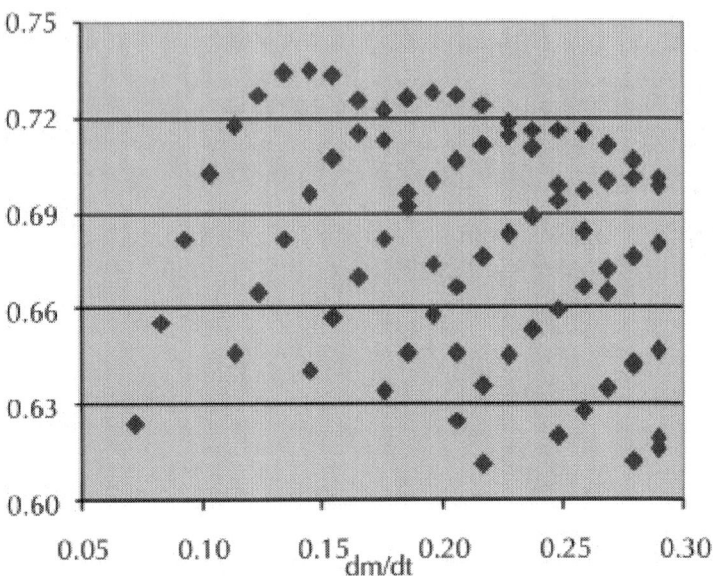

Figure 5: Compressor efficiency in function of compressor inlet mass flow rate.

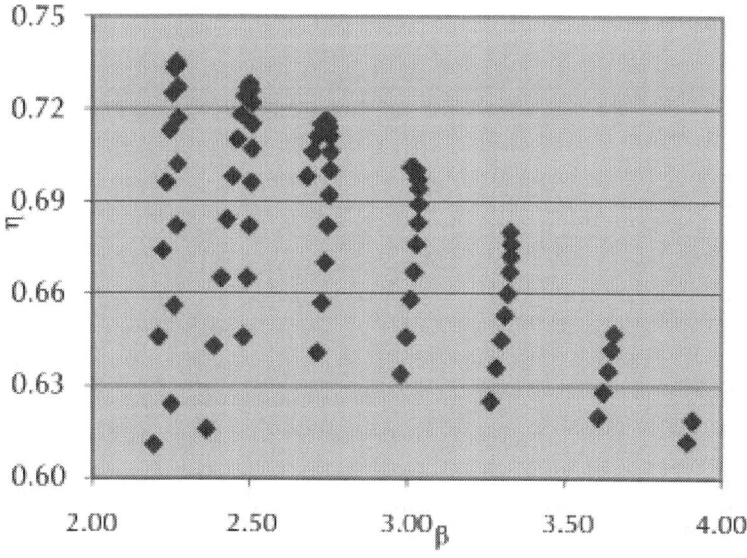

Figure 6: Compressor efficiency infunction of β.

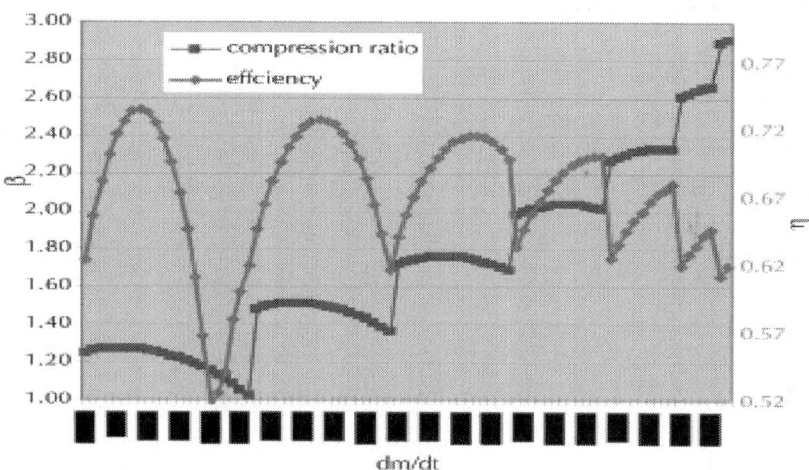

Figure 7: Compressor efficiency and b in function of compressor inlet mass flow rate.

Table 2: Tests main dadffta

Test	Rpm	Tex. (KJ)	Fuel cons. [l/min]	Test Elapsed time	# Tests
Steady	33,000	729	0.122	5m 50s	10
Steady	62.000	752	0.225	5m 40s	10
Steady	90,000	848	0.427	5m 42s	10
Steady	110,000	888	0.793	5m 20s	10
Variab.	33.000 - 108.000	727 - 888	max 0.797	5m 33s	4

Total tests time: 247m 32s = 4h 7m 12s.

bustion chamber CC, regenerator R, electrical motor EM, electrical generator EG, electronic control unit ECU, and case. External to the case, but of course properly connected to it, is the fuel tank assembly.

The operation of the UMGTG-UDR1 [5,6] is as follows:

- The compressor C pressurizes a stream of air;
- At the exit of compressor, the air stream is preheated in a regenerator;
- The preheated compressed air is then channelled into a combustor CC into which a suitable fuel is injected;
- The hot combustion gases are expanded through the turbine;
- The turbine powers the electric motor/generator (a dual effect engine);
- The particulars of the UMGTG device are:
- The generation of 500 - 2000 W or less of electrical power by means of a thermo-electrical conversion device of a sufficiently small size to be considered "portable";

- The internal arrangement of the device, in which a ultra-micro gas turbine plant provides the energy conversion from fuel to electrical power.

The packaging of said device is composed of several operative zones, each one designed to achieve the best possible performance and also designed in such a way that it might be built using commercially available components. The scope of the research is to provide electric power in a standalone operational mode, with limited fuel consumption and using a very compact device with a total length 0.334 m, width 0.184 m and height of 0.190 m and an occupied volume of 0.012 m³. The prototype and the proposal assembled device are shown in figure 8.

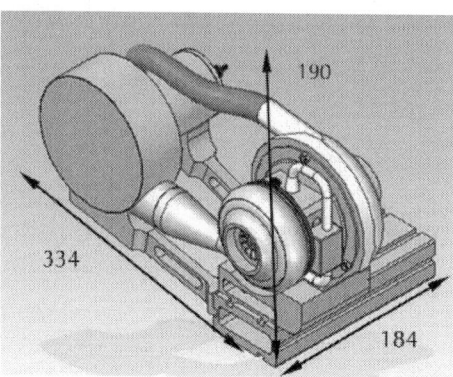

Figure 8: The proposed device and the prototype assembled at laboratory. The indicated dimensions are in mm.

It can be noticed that regeneration takes place inside the combustion chamber, thanks its particular architecture that will be discussed as follows. In Figure 9 the schematic representation of the UMGTG UDR1 cycle.

The Combustion Chamber Tests

As explained in the previous sub-section, the combustion chamber is an original design by the Mechanical and Aerospace Engineering Department of the University of Roma 1 Sapienza. A first order thermo-fluid-dynamic simulation has been carried out [7], and a more detailed study is in progress, to assess the attainment of a satisfactory thermal field for the liner and the completeness of combustion. A series of tests [5,6] provided the temperature map of the CC for a more exact analysis of the thermal flows and for experimentally validating the CFD results. A hole drilled on the top surface of the combustion chamber is the bay for a spark plug to activate the ignition of the (non pre-mixed) air-fuel mixture. On the cylindrical external surface, the second hole from the bottom is the compressor air inlet (B) the air flows through the spiralling duct and is pre-heated, and enters the combustion chamber from the upper side. The combustion air is injected with an intense swirl motion, in order to enhance mixing with the fuel that is in turn injected from a second port in the upper half of the cylinder. The combustion progresses and the exhausted gases exit from a port located on the opposite (bottom) side of the chamber and before being exhausted flow through a cylindrical shell around the spiralling air inlet to provide further preheating. Further developments of the combustion chamber are being explored, and include several possibilities: from an increased number of spiralling turns of a smaller cross area, to grooved internal surfaces. The scope of such improvements is to enhance the heat exchange and achieve a lower fuel consumption and a higher overall efficiency [8]. The option of inserting a secondary air pre-heater (thus implementing in effect a two-zone combustion) is also being considered. The inner diameter of the assembled combustion chamber is 22 mm, its outer diameter 42 mm and the overall height is 130 mm (Figure 10)

A fundamental step in the test of the combustion chamber is the temperature mapping of its surface. In particular, the measurements have been carried out at the following points (Figure 11):

- Inlet compressor air in the preheating zone;
- External wall of the inner cylinder;
- External wall of the outer cylinder;
- Exhaust gas volute wall.

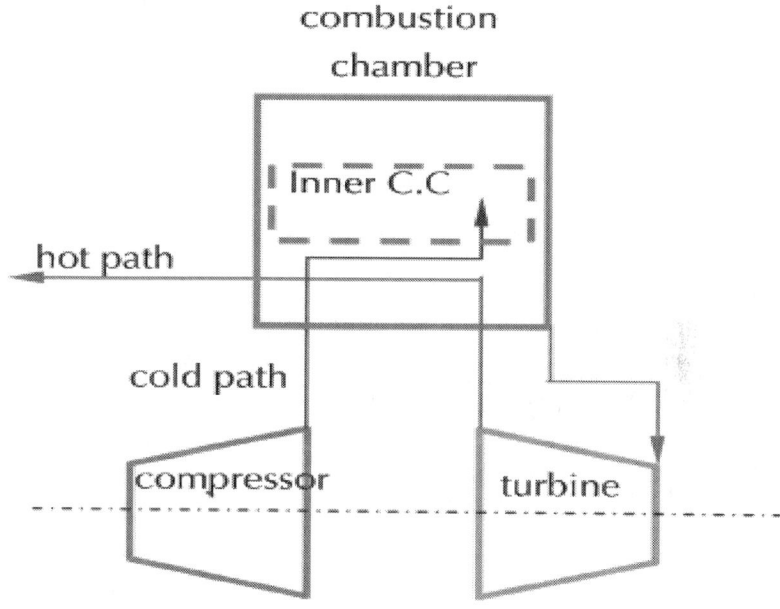

Figure 9: Schematic of UMGTGregenerated cycle.

Figure 10: The tested prototype of the combustion chamber.

The first part of the tests was devoted to the evaluation of the ignition time. Priming the combustion chamber is quite difficult when the inner chamber is still cold, and it required about 30 seconds. Once the chamber has been heated, the combustion is almost instantaneous. The following step has been the verification of combustion stability. These tests have been repeated with an interval between successive ignitions of three to five minutes. During the tests, attempts were made to regulate the air/fuel mixture in such a way as to establish different combustion regimes. In particular, after a first warm-up transient, a "complete" combustion was attained, with a typical blue colour of the flame that indicates the absence of unburned fractions. This combustion regime is maintained for about 10 - 15 minutes (see Table 3). We notice that, at the minimal variation of the fuel flow rate or inlet air, the flame presents unburned fractions again. The tests have underlined how the stability is obtained but the efficiency changes rapidly, when varying the air to fuel ratio.

Tests Results

After the ignition protocol was satisfactorily validated, the next series of tests were devoted to the measurement of the temperature at points A, B, C, D, E, F on the external wall of the chamber:

- In section D (see Figure 11), the values of temperature are comprised between 1173 K and 1288 K. The discrepancies are assumed to be due to the flow unsteadiness. A consistent correlation was observed between higher temperatures and higher combustion

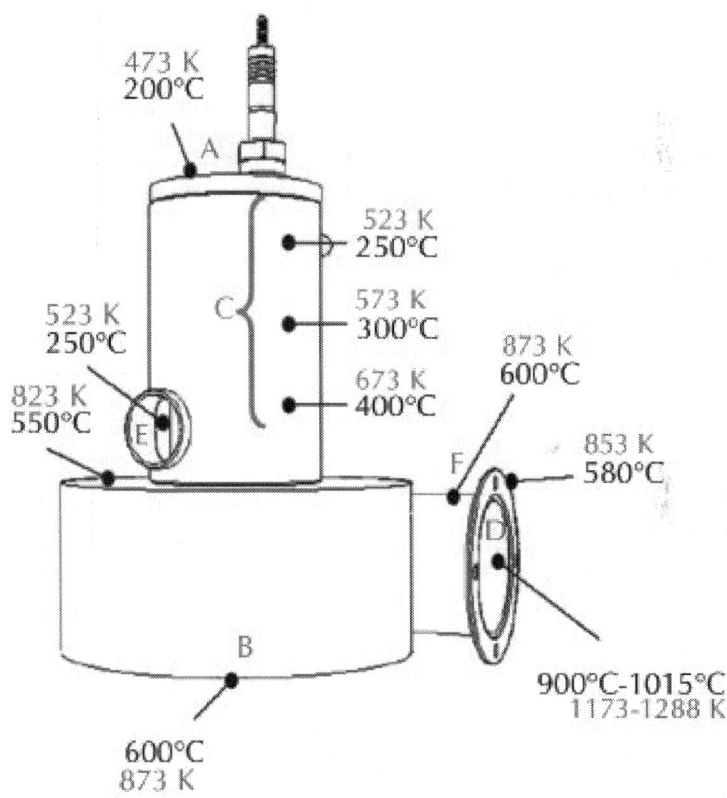

Figure 11: Sketch of the flows in the CC.

Table 3: Ignition time

CC conditions	Ignition time
Cold	30 s
Warm	instantaneous

deficiency conditions. In particular the temperature of 1288 K was measured keeping the K-thermocouple directly immersed in the exhaust gas (blue flame) for approximately 1 minute;

- On the external chamber wall (surface C in Figure 11), the measured temperatures varied between 473 K (200°C) on the top cover (cold zone) and 873 K (600°C) on the bottom one (hot zone). The cooler compressor air that flushes the preheating section exert a moderating effect on the wall temperature;

- The temperature measurement on the surface of the inner cylinder can be only performed in the immediate vicinity of the compressor air inlet section (point E, Figure 11). The amount of preheating obtained (a temperature increase of about 400 K) is reasonable in view of the following considerations:

- The temperature is relative to a wall on which "cold" (~310 K) compressed air impinges;

- The wall temperature of the operational machine is higher;

- The air "cup" temperature in the entry portion of the channel is lower than the average wall temperature of the inner cylinder.

- The above measurements confirm that the CC attains a regeneration degree (at least) equal to the one assumed in the preliminary design.

- Moreover, in the design phase a TIT of 1300 K was assumed, and the tests have measured a TIT of 1288 K, for all practical purposes sufficiently close to the design value, especially discounting the rather rough experimental conditions. These considerations lead to the conclusion that the design of CC was correct, in that the device operates at its foreseen design point. The results of the thermal test of the chamber are reported in Table 4.

TESTS MEASURE PROTOCOL

In this Section the testing procedures has been described. The campaign has been carried out, adopting the following procedures:

Definition of the Acquisition Time

This parameter has been defined according to the type of the required measures. It has been focused on 3 operational mode:

- Starter;
- Transitory;
- Steady condition.

    For each mode it has been chosen the "characteristic" time of acquisition, reported in Table 5.

Frequency of Acquisition

Such frequency has been varied within the sensitivity range of the instrument. The higher value is needed to acquire the higher frequencies in the transitory. So we adopt 3 working frequency for each operational mode:

Table 4: Combustion chamber thermal field

Position		Temperature (K]
A	Top cover	473
B	Bottom cover	873
C	External surface	523 - 673
D	Outlet gases	1173 -1288
E	Inlet air piping 5)3 (after 5 minutes of operation)	523
F	External outlet piping	873

Table 5: acquisition time

Operational mode	Acquisition time [s]	
	C.C.	Jetcat
Start up	60 - 120	30
Transients	180 - 300	30
Steady state	600 - 900	600

- 10 Hz (steady state);
- 100 Hz (start);
- 1000 Hz (transients).

Environment Parameters

The external environment temperature has been measured, regularly, every 15 minutes during the tests. At the scope a bulb thermometer has been used (sensibility of 0.5°C, error of ±0.2°C).

Instruments Nameplate

- Thermocouple TC9M-A-1-K-C-L-K for high temperature:

It is a sensor with special coating. Temperature measuring range is within 100°C to 1600°C, with protection sheath of recrystallised alumina KER 710 (C-799) tube and special platinum thimble;

- Thermocouple TC5-m-1-K-5-F-E-B-3:

Typical Fe-Co device. Material AISI 16 stainless steel with measuring range of 0°C to 600°C;

- Pressure gauge type TK-E-1-E-BO3U-M-V:

Measuring range of 0 to 3 bar.

Pre Treatment and Normalization of the Data

All data has been collected via PC and then has been so processed:

- Gauss Analysis of the deviations: mean values, mode, median and standard deviation have been automatically calculated by the software of the used acquisition card;
- Filtering: each data that exceeded 3s has been individually controlled, and neglected like spur if possible acquisition errors were not found;
- Normalization: the values are standardized in order to control that their distribution followed a Gaussiantype;
- Presentation: for each measure, the relative standard deviation and the medium value has been supplied ("rms").

Tests Elapsed Time

Regarding the hot tests on the combustion chamber, the tests have been subdivided in several sessions, with effective actual time for every session between 2 - 3 hours.

In the same manner, the tests have been subdivided in several sessions, with effective actual time for every session between 4 - 5 hours.

FUTURE DEVELOPMENTS AND CONCLUSIONS

The CC has been tested separately and now the device has been assembled for the final tests that are going to be carried out. The second step will be the deep study of the ignition procedure, to compute and measure all turbomachinery characteristics for the UDR1 prototype and for the Jetcat GT. These final tests should indicate the future possible optimization action to produce the final device (for the UMGTG UDR1), ready for a pre-commercialization, while, for what concern the 5 kW Jetcat GT, the tests would indicate how to optimize the device to reduce the fuel consumption and increase the bearings lifetime. Meanwhile we want to underline the approach to realizing devices at these scales utilizes industry-derived—and rather well demonstrated—micromachining technology. The economic impact of these devices will be dependent on the performance levels and the manufacturing costs, both of which have yet to be proven. It is certainly possible, however, that ultra micro GT sets may one day be competitive with conventional ICE machines for what the installed kW cost is concerned.

Even at much higher costs, they would have already no competitors as compact power sources for portable electronics

REFERENCES

1.   A. H. Epstein "Millimeter-Scale, Mems Gas Turbine Engines," Proceedings of ASME Turbo Expo 2003 on Power for Land, Sea, and Air, Atlanta, 16-19 June 2003, 28 p.

2.   A. H. Epstein, S. D. Senturia, G. Anathasuresh, A. Ayon, K. Breuer, K.-S. Chen, F. F. Ehrich, G. Gauba, R. Ghodssi, C. Groshenry, S. A. Jacobson, J. H. Lang, C.-C. Lin, A. Mehra, J. O. Mur Miranda, S. Nagle, D. J. Orr, E. Piekos, M. A. Schmidt, G Shirley, S. M. Spearing, C. S. Tan, Y.-S. Tzeng and I. A. Waitz, "Power Mems and Microengines," International Conference on Solid State Sensors and Actuators, Chicago, 16-19 June 1997, pp. 753-756.

3.   M. Ishihama, Y. Sakai, K. Matsuzuki and T. Hikone, "Structural Analysis of Rotating Parts of an Ultra Micro Gas Turbine," Proceedings of the International Gas Turbine Congress, Tokyo, 2-7 November 2003, 4 p.

4.   J. Peirs, T. Waumans, P. Vleugels, F. Al-Bender, T. Stevens, T. Verstraete, S. Stevens, R. D'hulsts, D. Verstraete, R. Van den Braembussche, J. Driesen, R. Puers, P. Hendrick, M. Baelmans and D. Reynaerts, "Micro Power Generation Based on Micro Gas Turbines: A Challenge," Proceedings of Institution of Mechanical Engineers, Vol. 221, Part C, pp. 489-497.

5.   R. Capata and E. Sciubba, "Further Development and Preliminary Testing of the α—Prototype of an Ultra-Micro Gas Turbine for Portable Power Generation," Proceedings of IMECE 2008 Conference, Boston, 2-7 November 2008.

6.   R. Capata and E. Sciubba, "Design and Performance Prediction of a Ultra-Micro Gas Turbine for Portable Power Generation," Proceedings of IMECE 2007 Conference, Seattle, 11-16 November 2007.

7.   G. S. Caveeiro, "Ultra Micro Gas Turbines Analysis," M. Thesis, Sapienza—University of Roma, Roma, 2006.

8.   Y. Wang, M. Wu, R. A. Yetter and V. Yang, "An Integrated Experimental and Numerical Study of Meso Scale Vortex Combustor Dynamics," 43rd AIAA Aerospace Sciences Meeting and Exhibit, Reno, 10-13 January 2005. http://arc.aiaa.org/doi/abs/10.2514/6.2005-941

A Fault Diagnosis Approach for Gas Turbine Exhaust Gas Temperature Based on Fuzzy C-means Clustering and Support Vector Machine

Zhi-tao Wang[1], Ning-bo Zhao[1], Wei-ying Wang[1, 2], Rui Tang[2], and Shu-ying Li[1]

[1]College of Power and Energy Engineering, Harbin Engineering University, Harbin 150001, China

[2]Harbin Marine Boiler & Turbine Research Institute, Harbin 150078, China

ABSTRACT

As an important gas path performance parameter of gas turbine, exhaust gas temperature (EGT) can represent the thermal health condition of gas turbine. In order to monitor and diagnose the EGT effectively, a fusion approach based on fuzzy C-means (FCM) clustering algorithm and support vector machine (SVM) classification model is proposed in this paper. Considering the distribution characteristics of gas turbine EGT, FCM clustering algorithm is used to realize clustering analysis and obtain the state pattern, on the basis of which the preclassification of EGT is completed. Then, SVM multiclassification model is designed to carry out the state pattern recognition and fault diagnosis. As an example, the historical monitoring data of EGT from an industrial gas turbine is analyzed and used to verify the performance of the fusion fault diagnosis approach presented in this paper. The results show that this approach can make full use of the unsupervised feature extraction ability of FCM clustering algorithm and the sample classification generalization properties of SVM multiclassification model, which offers an effective way to realize the online condition recognition and fault diagnosis of gas turbine EGT.

INTRODUCTION

With the development of high efficiency and clean energy, gas turbine plays an increasingly significant role in different domains, such as aviation and marine propulsion systems, electric power stations, and natural gas transportation petroleum [1]. With the increasing demand of security operation for gas turbine, the traditional regular maintenance technology has been unable to fully keep up with the actual demand and gas turbine health management technology has gradually become one of the most problems concerned by researchers and users in recent years [2]. In order to guarantee the gas turbine to run efficiently under the safe reliable condition, many sensors are often used to monitor the health state of gas turbine in the practical application. Massive amounts of data gathered by these sensors are easy to make difficulties in data analysis and affect the maintenance decision. Therefore, the choices of appropriate monitor parameters, signal processing methods and data mining techniques are very important to realize the health management of gas turbine.

Exhaust gas temperature (EGT) is an important gas path performance parameter of gas turbine, which can represent the thermal health condition of gas turbine [3, 4]. Considering the characteristics of different gas path performance monitoring parameters, the multiple linear regression models for analyzing the relationship between EGT and other parameters were established by Song et al. [5]. Their results showed that there were strong linear correlations between different gas path performance parameters and all the low turbine outlet pressure, high rotational speed, high pressure compressor outlet temperature, low rotational speed, and high pressure compressor outlet pressure could be reflected through the change of EGT. Yilmaz [6] also found the similar results by analyzing the relationship between EGT and other engine operational parameters at two different power settings, including maximum continuous and take-off, in the CFM56-7B turbofan engine. Hence, EGT is often used as an important parameter to evaluate the health state of gas turbine and determine the maintenance policy [7].

In the past half century, different methods have been developed to monitor and diagnose the EGT of gas turbine. Wang and Yang [8] analyzed many faults of PG6551B industrial gas turbine, such as turbine ablation, combustion component, and fuel system failure. They found that the uniformity of EGT could effectively reflect the feature of above fault. Chen et al. [9] proposed a general regression neural network (GRNN) approach to construct an autodetection network for EGT sensors, on the basis that they also studied the optimizing design of network and error controlling and developed the method of threshold for sensor detection. Based on the advantage of artificial neural networks (ANN), Muthuraman et al. [10] developed an autoassociative neural network approach to detect combustor-related damage by monitoring EGT. Błachnio and Pawlak [11] established a nonlinear observer and chose EGT as the important parameter to evaluate the health state of turbine blades. Korczewski [12–14] analyzed the change rules of EGT for a naval gas turbine engine under steady and unsteady operation conduction in detail. And they proposed an effective approach for detecting and evaluating the failures of the flow section and supply system of gas turbine by using EGT. Kenyon et al. [15] developed an intelligent system for detection of EGT anomalies in gas turbines by using the strong nonlinear mapping ability of ANN. Considering the characteristics of gas turbine operation control based on thermocouple measured exhaust temperatures, Xia et al. [16] discussed the application of Fiber-Bragg-grating-based sensing technology in the EGT measuring of gas turbine. Their results demonstrated that the fiber sensing method was more valuable for the monitoring and fault diagnosis of gas turbine because it could well reflect the changing of EGT. In order to increase the operational availability of industrial gas turbines, Yang et al. [17] presented a generalization of multidimensional linear regression to facilitate multisensor fault detection and signal reconstruction through the use of analytical optimization. Gülen et al. [18] discussed the relationship between EGT and other gas turbine performance parameters, on the basis of which an important diagnostic parameter named profile factor that was the ratio of the maximum exhaust thermocouple and the average of all exhaust temperature thermocouples was used to evaluate the performance of combustor and the whole gas turbine in their paper.

From the reviews discussed above, it is noted that the average EGT is often used to evaluate the health state of gas turbine in most researches. However, it may be more important and valuable to extract the relationship among different EGT sensors in order to realize condition monitor and fault diagnosis of gas turbine effectively. Although many studies have presented the effects of EGT distribution characteristics on health state of gas turbine or its hot sections and many analyses were discussed in detail, there was still a lack of systematic research in the area of online automatic identification and fault diagnosis for gas turbine EGT. Besides, EGT can be affected by many uncertain factors in the practical applications, which make it difficult to realize fault diagnosis quickly by using traditional model-driven approach. Therefore, it is very useful to develop a data-driven approach based on artificial intelligence technology in order to improve fault diagnosis accuracy.

The fault diagnosis of gas turbine EGT based on data-driven approach essentially is the cluster and classification of fault information. In the concrete implementation process, the training samples including normal and fault information need to be obtained firstly. Then the fault diagnosis model based on artificial intelligence algorithm can be established and trained by using training samples. The feature information collected from sensors will be inputted to the well trained fault diagnosis model and we can get the diagnostic results finally. Obviously, the establishment of fault information features space and design of artificial intelligence algorithm are two key steps to realize accurate fault diagnosis of gas turbine EGT. As mentioned above, many specific state patterns or fault types of gas turbine EGT cannot be determined directly based on experience in the practical applications due to the effects of many uncertain factors. Besides, enough prior knowledge including specific fault types is indispensable for the supervised artificial intelligence algorithms (such as ANN [19] and SVM [20]).

Considering the distribution characteristics of gas turbine EGT and the deficiencies of present literatures, a fusion approach based on FCM clustering algorithm and SVM classification model (FCM-SVM) is proposed in this paper. Firstly, FCM clustering algorithm is used to realize clustering analysis and obtain the state patterns of EGT, which means that the preclassification of EGT is completed. Then, SVM multiclassification model is designed to carry out the state pattern recognition and fault diagnosis of EGT. As an example, the historical monitoring data of EGT from an industrial gas turbine is analyzed to verify the effectiveness of the FCM-SVM approach finally.

The rest of this paper is organized as follows. In Section 2, the distribution characteristics of gas turbine EGT are described briefly. Section 3 introduces the basic theory of FCM clustering algorithm and SVM classification model in detail. The fusion fault diagnosis approach which combines FCM clustering with SVM is discussed in Section 4. Application examples and discussion are included in Section 5. Finally, Section 6 presents some conclusions.

SIGNAL FEATURE OF GAS TURBINE EGT

As mentioned above, it is very important to choose the appropriate measured parameters to monitor and diagnose the health state of gas turbine. Gas path and vibration parameters are two main types in the practical applications [2, 21]. Theoretically, as the most important gas path performance parameter for gas turbine, the outlet temperature of combustor chamber can not only affect the overall performance of engine, but also directly determine the ultimate strength of turbine blade. For example, the creep life of hot channel components can reduce the order of magnitude when the outlet temperature of combustor chamber increases 50°C [4], which may cause major fault and incur great maintenance costs. However, the outlet temperature of combustor chamber is usually so high that it cannot be measured directly by using conventional sensors. According to the well-defined Brayton thermodynamic cycle, there is a consistent relationship between the outlet temperature of combustor chamber and EGT. Therefore, EGT, as a measured parameter, is often used for gas engine control, condition monitoring, fault diagnosis, and maintenance decisions.

Compared with the average EGT, EGT profile can contain more information about the health state of gas turbine. Figures 1 and 2 show two EGT profiles with the same average EGT of an industrial gas turbine that has 12 EGT sensors. From Figures 1 and 2, it is easy to see that the EGT profiles of normal and fault condition are different although the average EGT are the same. This means that the fault information is incorrect or incomplete if only the average EGT is used to monitor the health state of gas turbine. Besides, Figures 1 and 2also show that all the sensors should give similar outputs when gas turbine operates in normal condition. If the component of gas turbine is failure, different temperatures will be observed. Therefore, the uniformity of EGT can more effectively reflect the health state of gas turbine, especially for steady state condition.

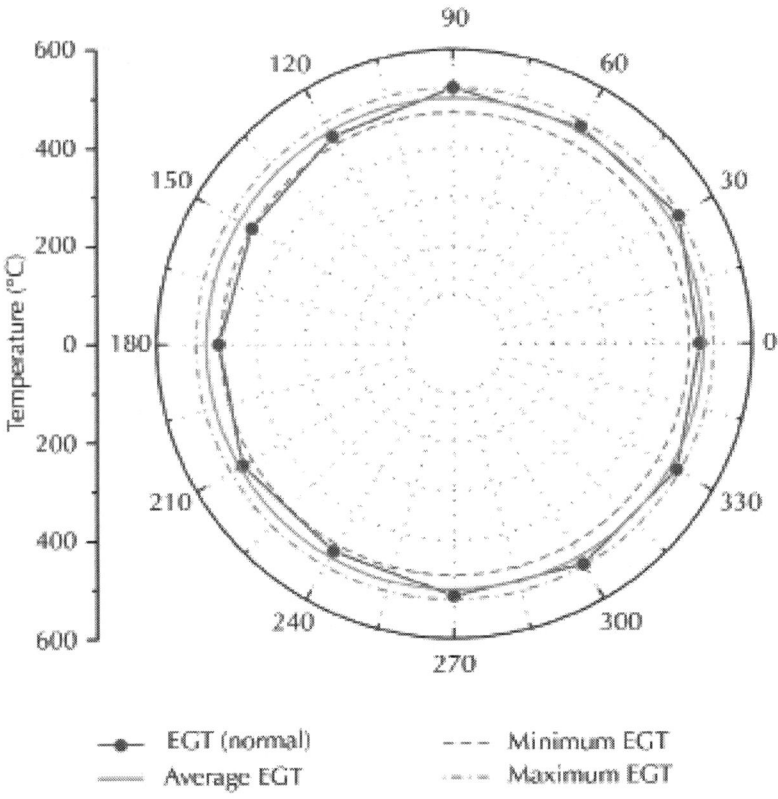

Figure 1: EGT profile of gas turbine with normal condition.

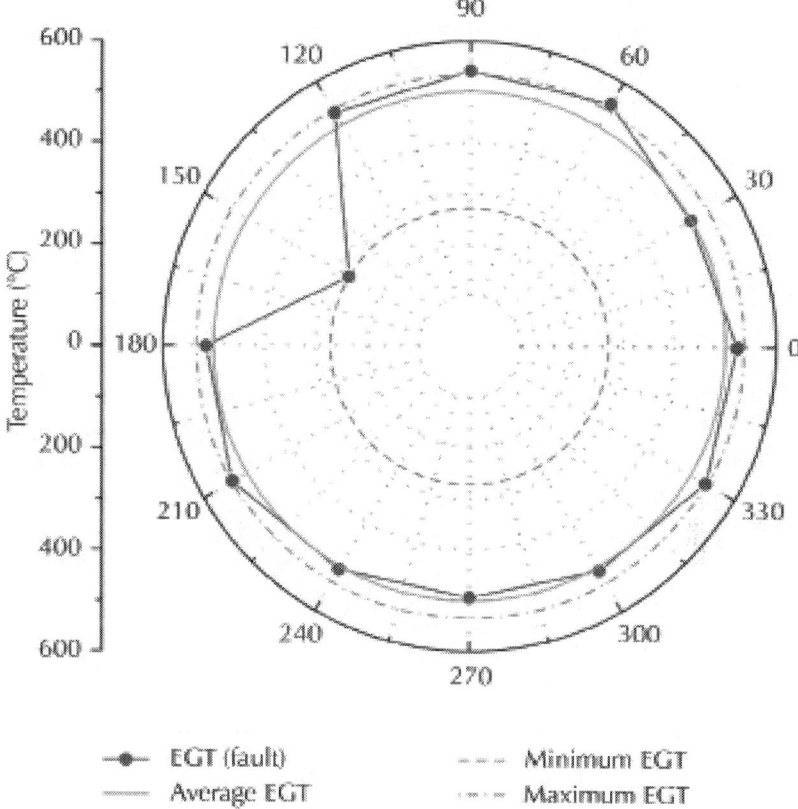

Figure 2: EGT profile of gas turbine with fault condition.

In order to quantitatively describe the uniformity of EGT, Mao [22] presented three indexes which can be calculated by the following functions. Assume that there are n sensors and their original outputs are T_i, i=1,......,n respectively:

$$H_1 = T_1' - T_n',$$

$$H_2 = T_1' - T_{n-1}',$$

$$H_3 = T_1' - T_{n-2}',$$

(1)

where T' is the transform value of and $T'_1 > T'_2 > \ldots\ldots > T'_{n-2} > T'_{n-1} > T'_n$
.

According to above indexes, it is obvious that all the values of H_1, H_2, and H_3 are smaller when gas turbine operates in normal condition. If a sensor fails, it usually causes H_1 or three indexes to increase. But the values of H_1 and H_2 or all three indexes can increase due to hot sections failure. However, it is worth pointing out that the above indexes only consider 4 EGT and others are ignored. Besides, it is difficult to diagnose the specific cause of failure when the sensor is fault because the above indexes ignore the adjacent information between different sensors. Therefore, there are some limitations to evaluate the uniformity of EGT only using the above three indexes. In order to solve this problem, all the measured EGT are used to realize cluster and fault diagnosis in this paper.

BASIC THEORY

Fuzzy C-means Clustering Algorithm

As an unsupervised machine learning method, FCM clustering algorithm was improved by Bezdek [23] in 1981 in order to solve the hard clustering problem by using fuzzy set theory. In the FCM clustering algorithm, membership degree function is used to indicate the extent to which each data point belongs to each cluster, and this information is also used to update the values of cluster centers [24]. Based on the concept of fuzzy C-partition, FCM clustering algorithm has been applied successfully in a wide variety of applications, such as image segmentation [25], data mining [26], thermal system monitoring [27], and fault diagnosis [28].

For the sample set $X=\{x_1,x_2,\ldots..,x_n\}$, the object of FCM clustering algorithm is to divide the sample set into groups and obtain the cluster centers by minimizing the following dissimilarity function [29]:

$$\min J_m(U,V) = \sum_{i=1}^{c}\sum_{k=1}^{n} u_{ik}^m d_{ik}^2,$$

(2)

where m is the fuzzy weighting parameter varying in the range $[1, \infty]$. The bigger the m, the more fuzzy the final cluster result. U is fuzzy partition matrix, V is cluster center matrix, and n and c are the number of samples and cluster centers, respectively. u_{ik} is the fuzzy membership degree of the kth sample in the ith cluster and it should be meeting the following three constraints [30]:

$$u_{ik} \in [0,1], \quad 1 \le i \le c, \ 1 \le k \le r$$

$$\sum_{i=1}^{c} u_{ik} = 1, \quad 1 \le k \le n,$$

$$\sum_{k=1}^{n} u_{ik} \in (0,n), \quad 1 \le i \le c.$$

$$(3)$$

For the distance d_{ik} between kth sample x_k and the centre of ith cluster V_i, it can be calculated by using Euclidean distance as follows:

$$d_{ik} = \|x_k - V_i\| = \sqrt{\sum_{j=1}^{m} \left(x_{kj} - V_{ij}\right)^2}, \quad 1 \le i \le c, \ 1 \le k \le n.$$

$$(4)$$

V_{ij} can be calculated by utilizing the following formulation:

$$V_{ij} = \frac{\sum_{k=1}^{n} u_{ik}^{m} x_{kj}}{\sum_{k=1}^{n} u_{ik}^{m}}, \quad 1 \le i \le c, \ 1 \le j \le m.$$

$$(5)$$

In essence, fuzzy cluster is performed through an iterative optimization by updating fuzzy membership degree [29]:

$$u_{ik}^{(s+1)} = \frac{1}{\sum_{j=1}^{c} \left(d_{ik}^{(s)} / d_{jk}^{(s)}\right)^{2/(m-1)}}, \quad 1 \le i \le c, \ 1 \le k \le n,$$

$$(6)$$

Where S is the iterative step.

When the below requirement is met, we can stop iteration and obtain the cluster result:

$$\left\| U^{(s+1)} - U^{(s)} \right\| \leq \varepsilon,$$

(7)

Where ε is the iterative threshold in the range [0, 1].

Based on above method, the cluster process of FCM clustering algorithm is virtually to determine the fuzzy membership degree and cluster centers through continuous iteration, which is shown in Figure 3.

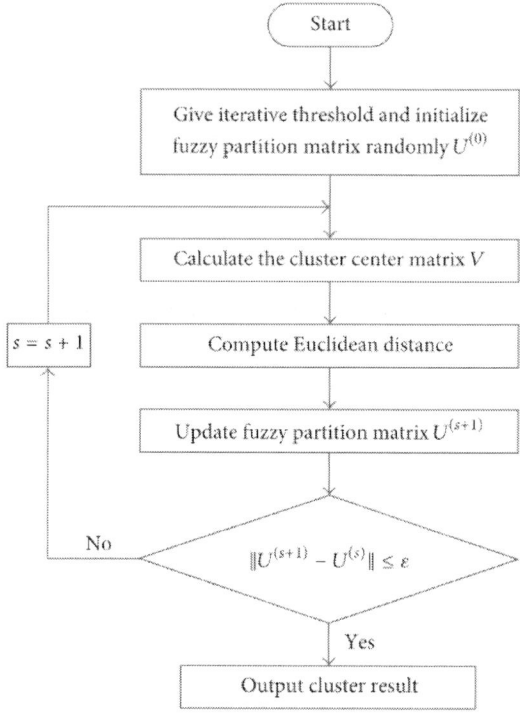

Figure 3: Calculation flowchart of fuzzy C-means clustering algorithm.

Support Vector Machine Classification Model

Compared with the conventional classifiers, support vector machine (SVM), developed by Vapnik [31], can effectively solve the classification problem by implementing the structure risk minimization based on statistical learning theory. Nowadays, SVM has been widely and successfully applied to detection and diagnosis of machine conditions due to its high accuracy and good generalization for a smaller number of samples [32, 33].

SVM is initially used to deal with binary classification problems. Its core idea is to transform the sample data from original space to a higher dimensional feature space through some nonlinear mapping functions and then find the optimal separating hyperplane in this feature space to realize linear classification. Figure 4 shows the classification principle based on SVM for the nonlinear classification problem.

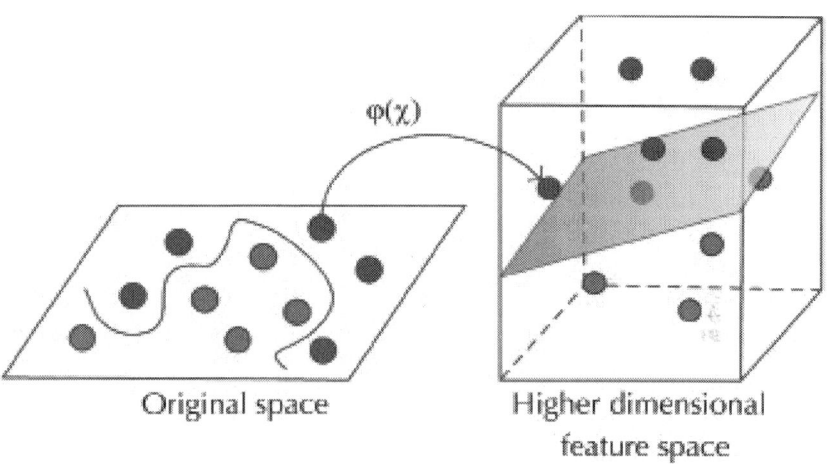

$\varphi(\chi)$

Original space   Higher dimensional feature space

Figure 4: Classification of two classes using SVM.

For the nonlinear training sample data set including two classes $\{x_i, y_i\}$, $i = 1, 2, \ldots, n$, $x \in R^n$, $y \in \{-1, +1\}$, n is the number of samples. The nonlinear mapping function $\varphi(x)$ will be used to transform the sample data from original space to a higher dimensional feature space and the optimal separating hyperplane can be constructed to maximize the margin between the two classes by the following linear function:

$$f(x) = [w \cdot \varphi(x)] + b, \tag{8}$$

where ω is the normal vector of optimal separating hyperplane and b is a scalar.

In essence, the solution of optimal separating hyperplane is the corresponding constraint optimization problem:

$$\min \quad \frac{1}{2}\|\omega\|^2 + C\sum_{i=1}^{n}\zeta_i$$

$$\text{s.t.} \quad y_i\left[w \cdot \varphi(x_i) + b\right] + \zeta_i \geq 1, \quad 1 \leq i \leq n$$

$$\zeta_i \geq 0, \quad 1 \leq i \leq n, \tag{9}$$

where C is the penalty factor that can realize the trade-off between empirical risk and confidence interval. z_i is slack factor.

Combining the method of Lagrange multipliers, the above convex optimization problem can be simplified into the dual quadratic optimization problem:

$$\max \quad L(a) = \sum_{i=1}^{n}a_i - \frac{1}{2}\sum_{i,j=1}^{n}a_i a_j y_i y_j \varphi(x_i) \cdot \varphi(x_j)$$

$$\text{s.t.} \quad a_i \geq 0, \quad 1 \leq i \leq n$$

$$\sum_{i=1}^{n}a_i y_i = 0, \tag{10}$$

where α is Lagrangian multiplier.

Then, the nonlinear decision function is described as

$$f(x) = \text{sign}\left(\sum_{i,j=1}^{n}a_i y_i\left(\varphi(x_i) \cdot \varphi(x_j)\right) + b\right). \tag{11}$$

In order to calculate the value of $\varphi(x_i)$. $\varphi(x_j)$, the kernel function $k(x_i, x_j)$ is used and the above function can be expressed as

$$f(x) = \text{sign}\left(\sum_{i,j=1}^{n} a_i y_i K\left(x_i, x_j\right) + b \right).$$

(12)

For the SVM, there are many kinds of kernel function, such as linear kernel, polynomial kernel, polynomial kernel, and radial basis function (RBF) kernel. Compared with other kernel functions, the RBF kernel can obtain the higher classification accuracy in many practical applications [34]. Therefore, the RBF kernel is used in this study.

As previously mentioned, SVM is initially designed for binary classification. However, there are often many faults in the practical applications, which mean that it is necessary to develop a method to deal with a multiclassification problem. Currently, different methods have been developed for the multiclassification based on SVM, such as "one-against-one," "one-against-all," and directed acyclic graph (DAG). According to the comparison results obtained by Hsu and Lin [35], the "one-against-one" method is more suitable for practical use than other methods. For the sample set including c class, c(c-1)/2 SVM classifiers can be constructed by using "one-against-one" method and every SVM classifier is trained.

FUSION FAULT DIAGNOSIS MODELING OF EGT BASED ON FCM-SVM APPROACH

As the supervised artificial intelligence method, enough prior knowledge including specific fault types is necessary for SVM classification model. However, the fault types of gas turbine EGT cannot be determined directly based on experience in the practical applications due to the effects of many uncertain factors. In order to achieve the automatic monitor and diagnosis of EGT effectively, a fusion approach based on FCM clustering algorithm and SVM classification model (FCM-

SVM) is proposed. Firstly, FCM clustering algorithm is used to realize clustering analysis and obtain the state patterns, which means that the preclassification of EGT is completed. Then, SVM multiclassification model is designed and used to carry out the online state pattern recognition and fault diagnosis of gas turbine EGT.

Figure 5 shows the fusion fault diagnosis framework of gas turbine EGT based on FCM-SVM approach. The detailed modeling processes are as follows.

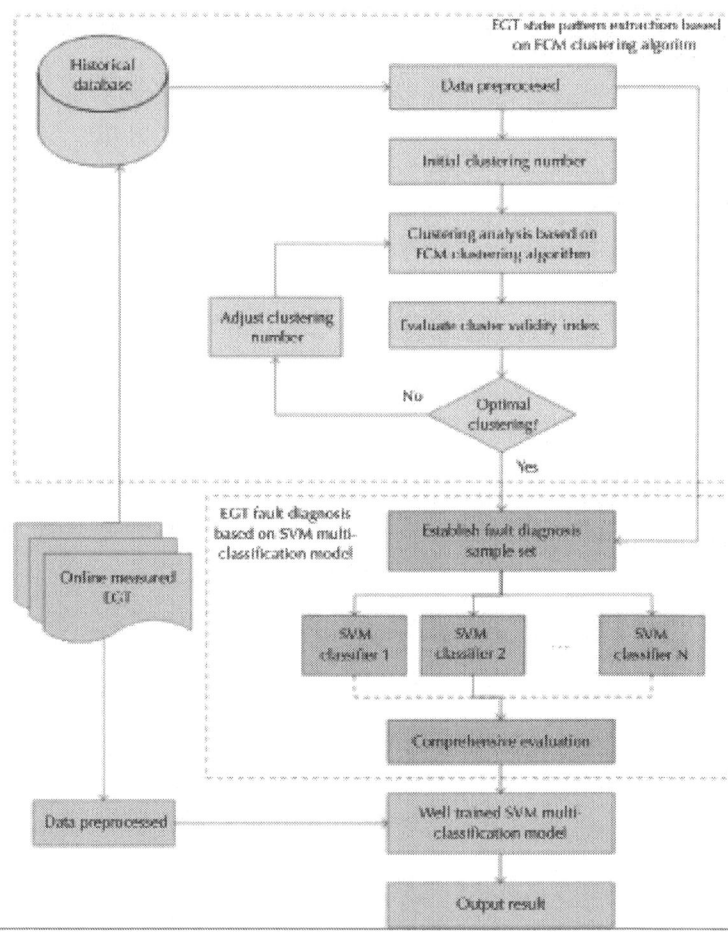

Figure 5: Fusion fault diagnosis framework of gas turbine EGT based on FCM-SVM approach.

Step 1: Generated sufficient EGT samples from the historical database and the essential preprocessing upon EGT data are carried out before data analysis, such as supplementary data, eliminating noise and outliers.

Step 2: According to the cluster process which is shown in Figure 3, FCM clustering algorithm is used to obtain the initial clustering results of gas turbine EGT.

Step 3: Cluster validity index λ(c) is used to evaluate the validity of clustering and determine the number of clusters. The c is optimum when λ(c) reaches its maximum value:

$$\lambda(c) = \frac{\sum_{i=1}^{c} \left(\sum_{k=1}^{n} u_{ik}^{m} \right) \left\| V_i - \overline{x} \right\|^2 / (c-1)}{\sum_{i=1}^{c} \sum_{k=1}^{n} u_{ik}^{m} \left\| x_k - V_i \right\|^2 / (n-c)},$$

(13)

$$\overline{x} = \frac{\sum_{i=1}^{c} \sum_{k=1}^{n} u_{ik}^{m} x_k}{n}.$$

(14)

Step 4: After obtaining the optimal clustering results, the fault diagnosis sample set including specific fault types can be established.

Step 5: SVM multiclassification model will be designed based on "one-against-one" method and trained by using fault diagnosis sample set.

Step 6: The measured EGT obtained from real gas turbine are preprocessed and inputted to the well trained SVM multiclassification model. Then we can get the final diagnostic results.

Step 7: The measured EGT also are stored into the historical database and used for later analysis.

CASE STUDY AND DISCUSSIONS

In order to demonstrate the effectiveness of FCM-SVM approach introduced in this paper, the historical monitoring data of EGT from one industrial single shaft gas turbine will be analyzed as a case study in this section.

Sample Data

As an industrial single shaft gas turbine, Taurus70 is made in solar turbines incorporated and used for power generation. 12 thermocouple temperature sensors are used to measure the EGT and the average EGT is about 505°C when gas turbine operates in a normal state.

Figure 6 shows the changing curves of 12 EGT varying with time under normal running state of gas turbine. And the EGT profiles can be seen in Figure 7. From Figures 6 and 7, it is clear that there is significant difference between the measured outputs of different thermocouple temperature sensors at the same time even when the gas turbine is running in a normal state. Therefore, much feature information will be ignored which can decrease the fault diagnosis accuracy if only the average EGT is used to evaluate and analyse the health state of EGT. Considering the operating conditions of gas turbine, 490-group data including 4 classes are taken to establish the original sample set. 470 samples are selected randomly as training samples and the remaining 20 samples are selected as testing samples.

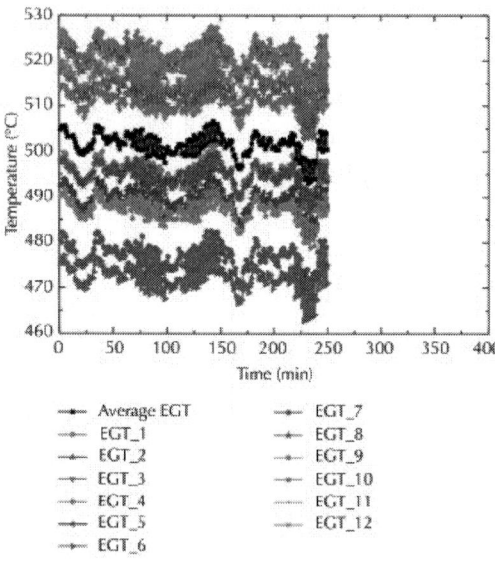

Figure 6: Real-time measured EGT curve of gas turbine with normal condition.

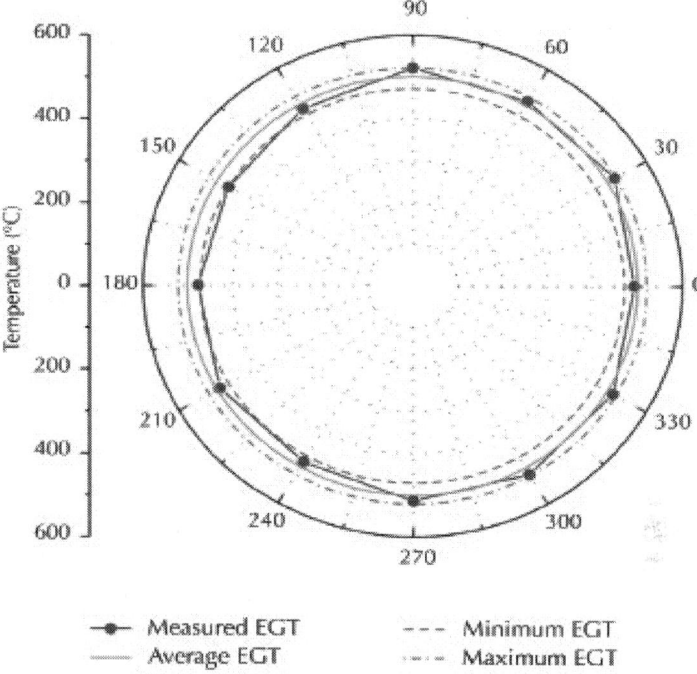

Figure 7: Real-time measured EGT profile of gas turbine with normal condition.

Optimal Clustering of EGT Based on FCM Clustering Algorithm

For the FCM clustering algorithm, it is very important to determine an appropriate number of clusters, which is called cluster validity problem. In this study, the cluster number is decided automatically by using the introduced cluster validity index which is shown in (13). Considering the computation complexity and accuracy, the scope of the number of clusters is commonly [2, $\sqrt{n}$] in practical process and n is the number of samples. For the 470 training samples shown in Table 1, the FCM clustering algorithm can stepwise iterate from 2 to 21 clusters. Figure 8 shows the changing trends of cluster validity index $\lambda(c)$ as a function of the number of clusters. Based on the result shown in Figure 8, it is clear that $\lambda(c)$ increases at first and then drops down with the increases of the number of clusters. And λ(c) can reach its maximum value when the number of clusters is 4, which is in agreement with the real samples class. Therefore, it is concluded that the FCM clustering algorithm is suitable for optimal clustering of gas turbine EGT.

Table 1: The original sample set of gas turbine EGT

Number	T1/°C	T2/°C	T3/°C	T4/°C	T5/°C	T6/°C	T7/°C	T8/°C	T9/°C	T10/°C	T11/°C	T12/°C
1	493.6847	519.3781	510.2112	521.7342	488.1656	471.3032	475.5255	493.422	486.8506	513.879	518.3308	514.9266
2	497.8859	524.3516	513.879	524.09	493.422	475.5255	480.0065	497.8859	489.743	517.5453	523.043	519.1163
3	496.3108	521.996	512.3073	524.09	490.5316	473.151	478.9526	497.0984	488.1656	515.7123	520.9489	516.7598
4	500.2478	524.6133	513.355	526.9688	493.6847	477.1075	481.8502	499.723	491.8456	519.1163	523.043	520.6871
5	497.8859	521.7342	511.7833	522.7812	491.3201	473.9427	478.6891	496.5733	488.9544	515.9742	520.6871	516.7598
6	499.723	524.8751	514.1408	526.4454	494.21	476.3166	481.3235	499.1981	492.1084	518.8545	523.8282	520.1635
7	495.7857	520.6871	512.3073	523.3047	490.7945	473.4149	477.8983	495.2605	488.9544	514.1408	520.9489	516.7598
8	499.4606	525.1368	513.0931	526.4454	493.422	476.0529	481.5869	498.4109	491.5829	518.069	522.5195	519.3781
9	496.5733	523.043	513.355	524.09	491.8456	475.5255	480.5334	497.8859	489.2173	517.2835	521.7342	518.069
10	497.8859	523.043	513.879	525.1368	492.6339	475.2617	480.5334	497.6234	490.2688	517.2835	523.5664	519.9018
:	:	:	:	:	:	:	:	:	:	:	:	:
486	498.6733	525.3986	514.9266	525.922	493.9474	476.3166	482.1135	498.1484	490.5316	518.5927	522.7812	519.1163
487	498.6733	523.5664	514.9266	526.7071	493.1593	476.5803	482.1135	497.6234	490.7945	518.5927	523.043	519.64
488	498.6733	523.043	512.5692	524.8751	492.6339	475.7892	480.0065	497.0984	488.9544	518.069	521.7342	518.069
489	498.4109	522.5195	513.0931	524.09	491.8456	474.9979	480.2699	497.3609	489.4802	517.0217	521.2107	518.8545
490	499.4606	523.3047	514.6647	525.6602	492.6339	475.7892	481.0601	498.1484	491.0573	517.0217	521.7342	519.9018

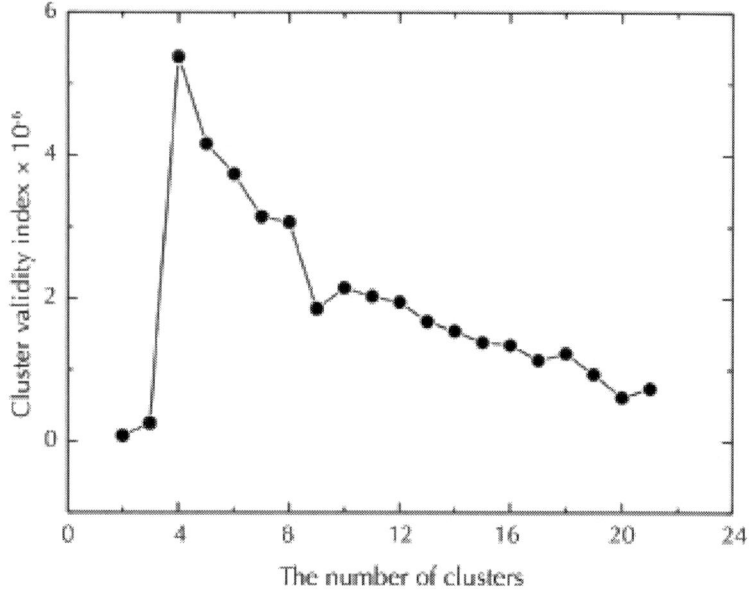

Figure 8: The effect of the number of clusters on cluster validity index λ(c).

Considering the high dimension characteristics of samples, it is difficult to realize graphical analysis directly. In this paper, three temperature uniformity indexes described by Mao [22] are used to further analyze and evaluate the cluster results of gas turbine EGT by using graphic approach. Figures 9–11 show the cluster results of gas turbine EGT based on FCM clustering algorithm. It may be clearly observed in Figures 9–11 that all the three temperature uniformity indexes of F1 class are relatively

small $\left(0°C \leq H_1 \leq 30°C, 0°C \leq H_2 \leq 30°C, \text{and } 0°C \leq H_3 \leq 30°C\right)$. This means that 12 thermocouple temperature sensors give the almost same outputs, which belongs to normal state. Compared with F1 class, F2 class has the following characteristics:

$40°C \leq H_1 \leq 70°C, 30°C \leq H_2 \leq 45°C, \text{and } 30°C \leq H_3 \leq 45°C$.   The actual experimental results show that the fundamental reason for this phenomenon is turbine blade wear which can cause a difference of enthalpy drop between different turbine blade passages. For the F3 class, all the three temperature uniformity indexes are very large

$\left(140°C \leq H_1, 110°C \leq H_2, \text{and } 110°C \leq H_3\right)$ due to the effects of gas turbine load rejection. In addition, a careful inspection of Figures 9 and 10 reveals that the temperature uniformity index H_1 is significantly larger than the

other two indexes $\left(500°C \leq H_1, 0°C \leq H_2 \leq 30°C, \text{and } 0°C \leq H_3 \leq 30°C\right)$. It means that one of the 12 thermocouple temperature sensors is fault which can result in a smaller output.

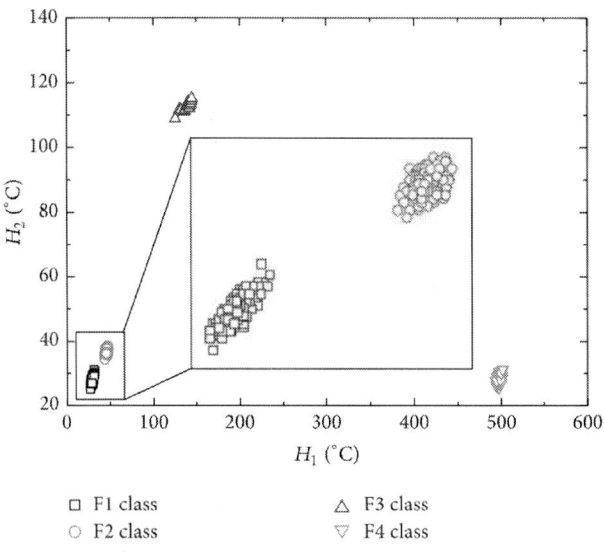

Figure 9: Graphic clustering result by using H_1 and H_2.

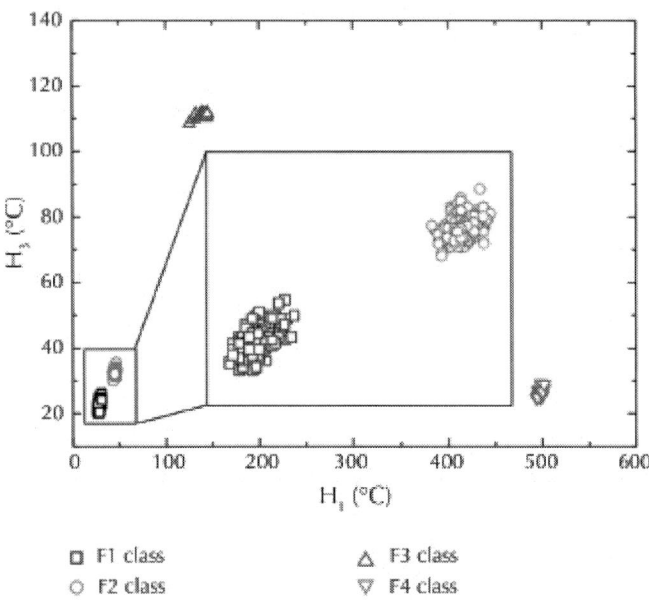

Figure 10: Graphic clustering result by using H_1 and H_3.

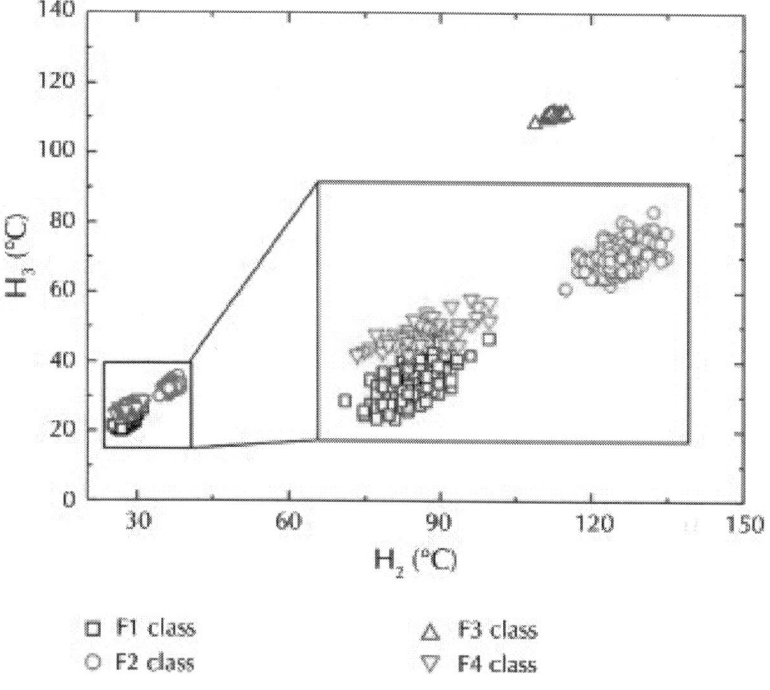

Figure 11: Graphic clustering result by using H_2 and H_3.

EGT Fault Diagnosis Based on SVM Classification Model

According to the optimal clustering results, the fault diagnosis training sample set including fault types can be established, which is shown in Table 2. Then on this basis, we can develop 6 SVM classifiers based on "one-against-one" method. Table 3 shows the fault diagnosis performance of SVM multiclassification model for the training samples. From Table 3, it is clear that the fault diagnosis accuracy rate of trained SVM multiclassification model is 100% for the training samples, which means that the SVM multiclassification model has been well trained for fault diagnosis of gas turbine EGT.

Table 2: Fault diagnosis training sample set of EGT

Type	T1/°C	T2/°C	T3/°C	T4/°C	T5/°C	T6/°C	T7/°C	T8/°C	T9/°C	T10/°C	T11/°C	T12/°C
	493.6847	519.3781	510.2112	521.7342	488.1656	471.3032	475.5255	493.422	486.8506	513.879	518.3308	514.9266
	497.8859	524.3516	513.879	524.09	493.422	475.5255	480.0065	497.8859	489.743	517.5453	523.043	519.1163
F1	496.3108	521.996	512.3073	524.09	490.5316	473.151	478.9526	497.0984	488.1656	515.7123	520.9489	516.7598
	500.2478	524.6133	513.355	526.9688	493.6847	477.1075	481.8502	499.723	491.8456	519.1163	523.043	520.6871
	:	:	:	:	:	:	:	:	:	:	:	:
	489.743	511.5213	502.3465	512.0453	481.0601	464.9604	474.2065	490.2688	480.7968	502.8711	511.2593	506.2799
	489.4802	512.0453	502.8711	513.617	482.1135	466.5472	473.9427	490.7945	482.1135	504.4445	512.5692	507.5906
F2	487.9026	510.9973	501.0349	512.3073	480.0065	465.2249	472.3592	490.2688	480.5334	502.3465	510.9973	506.5421
	488.4285	512.0453	503.6579	512.8312	480.7968	466.8116	473.151	490.7945	481.8502	502.8711	512.8312	507.5906
	:	:	:	:	:	:	:	:	:	:	:	:
	444.8924	448.348	487.1684	474.5367	461.6006	427.5448	244.4364	333.8674	236.1834	323.3517	431.559	419.7637
	444.0943	448.348	485.8542	472.955	461.3361	427.009	244.7409	334.1505	236.7968	322.4955	430.7567	420.3012
F3	444.8924	448.6136	485.8542	474.0095	461.0715	427.5448	244.7409	334.4335	236.1834	322.781	431.559	420.5699
	443.0299	448.0823	485.8542	473.2187	460.807	427.009	243.8271	333.8674	236.1834	321.9244	430.2217	420.3012
	:	:	:	:	:	:	:	:	:	:	:	:
	472.3592	463.3729	454.6267	468.3976	−17.7778	471.3032	481.3235	476.8439	456.2188	462.5787	468.926	479.2161
	472.3592	464.6959	454.3612	468.6618	−17.7778	471.0391	482.6401	477.1075	455.4228	462.0493	469.4544	477.3712
F4	470.511	461.5197	451.174	466.8116	−17.7778	469.4544	480.7968	474.2065	454.3612	460.4602	467.869	478.4255
	470.7751	462.314	451.7055	466.8116	−17.7778	469.7186	481.3235	475.5255	454.892	460.1953	467.869	477.1075
	:	:	:	:	:	:	:	:	:	:	:	:

Table 3: Fault diagnosis performance of SVM multiclassification model for training samples

Type	Number of training samples	Number of accurate diagnoses	Accuracy rate/%
F1	200	200	100
F2	200	200	100
F3	20	20	100
F4	50	50	100

Then the testing samples shown in Table 1 are used to further demonstrate the effectiveness of SVM multiclassification model. Table 4 shows the comparison between actual results and fault diagnosis results by using the well trained SVM multiclassification model for testing samples. Based on the results shown in Table4, it is demonstrated that the well trained SVM multiclassification model can effectively diagnose the fault of gas turbine EGT with a 95% accuracy rate for the testing samples. Besides, the reason of misclassification is that the sample data is obtained when the turbine blade wear or corrosion is not severe. In order to compare with other models, backpropagation

(BP) neural network model is also employed to make the same fault diagnosis and the results are also listed in Table 4. The comparative analysis shows that SVM classification model can improve the fault diagnosis accuracy of gas turbine EGT significantly compared with BP neural network model. All these indicate that SVM is more suitable for fault diagnosis of gas turbine EGT.

Table 4: The comparison results of different fault diagnosis models for testing samples

Number	SVM1	SVM2	SVM3	SVM4	SVM5	SVM6	SVM model	BP model	Actual results
1	F1	F1	F1	F2	F2	F3	F1	F1	F1
2	F1	F1	F1	F2	F2	F3	F1	F1	F1
3	F1	F1	F1	F2	F2	F3	F1	F1	F1
4	F1	F1	F1	F2	F2	F3	F1	F2	F1
5	F1	F1	F1	F2	F2	F3	F1	F1	F1
6	F2	F1	F1	F2	F2	F3	F2	F2	F2
7	F1	F1	F1	F2	F2	F3	F1	F1	F2
8	F2	F1	F1	F2	F2	F3	F2	F2	F2
9	F2	F1	F1	F2	F2	F3	F2	F2	F2
10	F2	F1	F1	F2	F2	F3	F2	F1	F2
11	F2	F3	F1	F3	F2	F3	F3	F3	F3
12	F2	F3	F1	F3	F2	F3	F3	F3	F3
13	F2	F3	F1	F3	F2	F3	F3	F3	F3
14	F2	F3	F1	F3	F2	F3	F3	F3	F3
15	F2	F3	F1	F3	F2	F3	F3	F3	F3
16	F1	F3	F4	F3	F4	F4	F4	F4	F4
17	F1	F3	F4	F3	F4	F4	F4	F4	F4
18	F1	F3	F4	F3	F4	F4	F4	F4	F4
19	F1	F3	F4	F3	F4	F4	F4	F4	F4
20	F1	F3	F4	F3	F4	F4	F4	F4	F4

CONCLUSIONS

Considering the distribution characteristics of gas turbine EGT and its effect on the health state of gas turbine, a fusion approach based on FCM clustering algorithm and SVM classification model (FCM-SVM) is proposed and successfully applied to an industrial gas turbine in this paper. In the analysis presented in this study, it is demonstrated that FCM-SVM based approach can make full use of the unsupervised feature extraction ability of FCM clustering algorithm and the sample classification generalization properties of SVM multiclassification model, which offers an effective way to realize the online condition recognition and fault diagnosis of gas turbine EGT. In the concrete implementation process, the introduced FCM clustering algorithm is a good alternative to achieve automatic identification of the fault types of gas turbine EGT. In other words, it is effective to overcome the influence of experience judgment on fault types. Besides, the introduction of SVM multiclassification model has a great potential to improve the fault diagnosis performance of gas turbine EGT. It is worth noticing that the study of this paper is only focused on researching the artificial intelligence approach for the condition recognition and fault diagnosis of gas turbine EGT but ignores the effects of many other parameters such as inlet temperature of gas turbine. Therefore, more studies and improvement about the application of this approach are needed further.

REFERENCES

1.   W. Y. Wang, Z. Q. Xu, R. Tang, S. Y. Li, and W. Wu, "Fault detection and diagnosis for gas turbines based on a kernelized information entropy model," The Scientific World Journal, vol. 2014, Article ID 617162, 13 pages, 2014. ·

2.   A. J. Volponi, "Gas turbine engine health management: past, present, and future trends," Journal of Engineering for Gas Turbines and Power, vol. 136, no. 5, Article ID 051201, 2014. · ·

3.   T. Palmé, F. Liard, and D. Therkorn, "Similarity based modeling for turbine exit temperature spread monitoring on gas turbines," in Proceedings of the ASME Turbo Expo 2013: Turbine Technical Conference and Exposition (GT ‹13), American Society of Mechanical Engineers, San Antonio, Tex, USA, June 2013. · ·

4.   P. E. Patrick Hamilton and D. Ha, "Exhaust gas temperature capabilities now in system 1 software," Product Update, vol. 25, no. 1, pp. 88–89, 2005.

5.   Y. X. Song, K. X. Zhang, and Y. S. Shi, "Research on aeroengine performance parameters forecast based on multiple linear

regression forecasting method," Journal of Aerospace Power, vol. 24, no. 2, pp. 427–431, 2009.

6.   I. Yilmaz, "Evaluation of the relationship between exhaust gas temperature and operational parameters in CFM56-7B engines," Proceedings of the Institution of Mechanical Engineers—Part G: Journal of Aerospace Engineering, vol. 223, no. 4, pp. 433–440, 2009. · ·

7.   E. M. He and L. T. Song, "Analysis of EGT and measures to increase the EGT margin," Aviation Engineering & Maintenance, vol. 6, pp. 20–21, 1999.

8.   X. F. Wang and J. M. Yang, "Analysis and treatment of larger exhaust dispersity fault for PG6551B gas turbine," Gas Turbine Technology, vol. 17, no. 2, pp. 58–61, 2004.

9.   J. Chen, Y. H. Wang, and S. L. Weng, "Application of general regression neural network in fault detection of exhaust temperature sensors on gas turbines," Proceedings of the Chinese Society of Electrical Engineering, vol. 29, no. 32, pp. 92–97, 2009.

10. S. Muthuraman, J. Twiddle, M. Singh, and N. Connolly, "Condition monitoring of SSE gas turbines using artificial neural networks," Insight: Non-Destructive Testing and Condition Monitoring, vol. 54, no. 8, pp. 436–439, 2012. · ·

11. J. Błachnio and W. I. Pawlak, "Damageability of gas turbine blades-evaluation of exhaust gas temperature in front of the turbine using a non-linear observer," in Advances in Gas Turbine Technology, chapter 19, pp. 435–464, 2011. ·

12. Z. Korczewski, "Exhaust gas temperature measurements in diagnostic examination of naval gas turbine engines," Polish Maritime Research, vol. 18, no. 2, pp. 37–43, 2011. · ·

13. Z. Korczewski, "Exhaust gas temperature measurements in diagnostic examination of naval gas turbine engines—part II: unsteady processes," Polish Maritime Research, vol. 18, no. 3, pp. 37–42, 2011.

14. Z. Korczewski, "Exhaust gas temperature measurements in diagnostic examination of naval gas turbine engines," Polish Maritime Research, vol. 18, no. 4, pp. 49–53, 2011. · ·

15. A. D. Kenyon, V. M. Catterson, and S. D. J. McArthur, "Development of an intelligent system for detection of exhaust gas temperature

anomalies in gas turbines," Insight: Non-Destructive Testing and Condition Monitoring, vol. 52, no. 8, pp. 419–423, 2010. · ·

16. H. Xia, D. Byrd, S. Dekate, and B. Lee, "High-density fiber optical sensor and instrumentation for gas turbine operation condition monitoring," Journal of Sensors, vol. 2013, Article ID 206738, 10 pages, 2013. · ·

17. Z. Yang, B. W. K. Ling, and C. Bingham, "Fault detection and signal reconstruction for increasing operational availability of industrial gas turbines," Measurement, vol. 46, no. 6, pp. 1938–1946, 2013. · ·

18. S. C. Gülen, P. R. Griffin, and S. Paolucci, "Real-time on-line performance diagnostics of heavy-duty industrial gas turbines," Journal of Engineering for Gas Turbines and Power, vol. 124, no. 4, pp. 910–921, 2002. · ·

19. D. H. Seo, T. S. Roh, and D. W. Choi, "Defect diagnostics of gas turbine engine using hybrid SVM-ANN with module system in off-design condition," Journal of Mechanical Science and Technology, vol. 23, no. 3, pp. 677–685, 2009. · ·

20. Y. Hao, J. G. Sun, G. Q. Yang, and J. Bai, "The application of support vector machines to gas turbine performance diagnosis," Chinese Journal of Aeronautics, vol. 18, no. 1, pp. 15–19, 2005. · ·

21. W. X. Wang and W. Z. An, "Engine vibration fault diagnosis research based on fuzzy clustering method of gas turbine," Gas Turbine Technology, vol. 26, no. 3, pp. 44–47, 2013.

22. H. J. Mao, "Analysis of exhaust temperature monitor and protection function for gas turbine," Huadian Technology, vol. 31, no. 8, pp. 11–15, 2009.

23. J. C. Bezdek, Pattern Recognition with Fuzzy Objective Function Algorithms, Kluwer Academic, New York, NY, USA, 1981.

24. S. Wikaisuksakul, "A multi-objective genetic algorithm with fuzzy c-means for automatic data clustering," Applied Soft Computing, vol. 24, pp. 679–691, 2014. ·

25. W. Cai, S. Chen, and D. Zhang, "Fast and robust fuzzy c-means clustering algorithms incorporating local information for image segmentation," Pattern Recognition, vol. 40, no. 3, pp. 825–838, 2007. · ·

26. S. N. Omkar, S. Suresh, T. R. Raghavendra, and V. Mani, "Acoustic emission signal classification using fuzzy C-means clustering," in Proceedings of the 9th International Conference on Neural Information Processing, vol. 4, pp. 1827–1831, 2002.

27. H. Zhao, P. H. Wang, J. Qian, Z. Su, and X. Peng, "Modeling for target-value of boiler monitoring parameters based on fuzzy C-means clustering algorithm," Proceedings of the Chinese Society of Electrical Engineering, vol. 31, no. 32, pp. 16–22, 2011.

28. C. Xu, P. Zhang, G. Ren, and J. Fu, "Engine wear fault diagnosis based on improved semi-supervised fuzzy c-means clustering," Journal of Mechanical Engineering, vol. 47, no. 17, pp. 55–60, 2011. · ·

29. H. B. Sahu, S. S. Mahapatra, and D. C. Panigrahi, "Fuzzy c-means clustering approach for classification of Indian coal seams with respect to their spontaneous combustion susceptibility," Fuel Processing Technology, vol. 104, pp. 115–120, 2012. · ·

30. N. R. Pal and J. C. Bezdek, "On cluster validity for the fuzzy c-means model," IEEE Transactions on Fuzzy Systems, vol. 3, no. 3, pp. 370–379, 1995. · ·

31. V. N. Vapnik, The Nature of Statistical Learning Theory, Springer, 1995. · ·

32. Q. Hu, Z. He, Z. Zhang, and Y. Zi, "Fault diagnosis of rotating machinery based on improved wavelet package transform and SVMs ensemble," Mechanical Systems and Signal Processing, vol. 21, no. 2, pp. 688–705, 2007. · ·

33. A. Widodo and B. S. Yang, "Support vector machine in machine condition monitoring and fault diagnosis," Mechanical Systems and Signal Processing, vol. 21, no. 6, pp. 2560–2574, 2007. · ·

34. J. Yang, Y. Zhang, and Y. Zhu, "Intelligent fault diagnosis of rolling element bearing based on SVMs and fractal dimension," Mechanical Systems and Signal Processing, vol. 21, no. 5, pp. 2012–2024, 2007. · ·

35. C. W. Hsu and C. J. Lin, "A comparison of methods for multiclass support vector machines," IEEE Transactions on Neural Networks, vol. 13, no. 2, pp. 415–425, 2002. · ·

4

Gas Turbine Blade Damper Optimization Methodology

R. K. Giridhar[1] P. V. Ramaiah[2] G. Krishnaiah[2], and S. G. Barad[1]

[1]Vibration Engineering Group, Gas Turbine Research Establishment, CV Raman Nagar, Bangalore 560093, India

[2]Mechanical Engineering Department, Sri Venkateswara University, Tirupati 517502, India

ABSTRACT

The friction damping concept is widely used to reduce resonance stresses in gas turbines. A friction damper has been designed for high pressure turbine stage of a turbojet engine. The objective of this work is to find out effectiveness of the damper while minimizing resonant stresses for sixth and ninth engine order excitation of first flexure mode. This paper presents a methodology that combines three essential phases of friction damping optimization in turbo-machinery. The first phase is to develop an analytical model of blade damper system. The second phase is experimentation and model tuning necessary for response studies while the third phase is evaluating damper performance. The reduced model of blade is developed corresponding to the mode under investigation incorporating the friction damper then the simulations were carried out to arrive at an optimum design point of the damper. Bench tests were carried out in two phases. Phase-1 deals with characterization of the blade-dynamically and the phase-2 deals with finding optimal normal load at which the blade resonating response is minimal for a given excitation. The test results are discussed, and are corroborated with simulated results, are in good agreement.

INTRODUCTION

The friction damping concept is widely applied in turbomachinery applications, especially at hot end parts, to reduce resonance stresses. A typical application of this in gas turbines. They are popularly called as "friction damper," "cottage-roof damper" or "under platform damper." This damper is loaded by centrifugal force against the underside of the platforms of two adjacent blades. The main design criterion for such devices is to determine the optimum damper configuration or the damper mass or both in order to reduce the dynamic stresses to maximum possible extent. For example, if the damper mass is too small for a given configuration, the friction force will not be large enough to dissipate sufficient energy. On the other hand, if the damper mass is too large, it will get into "stick" condition, thereby limiting the relative motion across the interface and hence the amount of energy dissipation. In both cases, the friction damper will be inefficient, and between these two extremes there exists an optima.

A good review of the friction damping concept in turbomachinery applications is given by Griffin [1]. Theoretical analysis and the optimization of this simple device is difficult because of marked nonlinearity and assumptions about the contact characteristics and damper behavior. Several friction damper models and analysis methods have been proposed in past. The simplest and most commonly used model reported in the literature is a macroslip contact model [2–4]. There are also several micro-slip friction models reported which are more appropriate in case of high normal loads [5–10]. The macroslip model under this high normal load shows damper in lockup condition.

Although significant advances are made in theoretical modeling of friction dampers and analysis methods, turbomachinary manufacturers still rely on previous experience and empirical data rather than computer-based predictions alone for friction damper optimization. This has been mainly due to the oversimplification introduced in the models regarding the basic contact behavior and/or damper geometry and the inability to analyze representative-size models due to excessive computational cost [4].

The latest and most advanced contact models make use of three parameters to characterize the contact behavior, namely, friction coefficient, tangential contact stiffness and normal contact stiffness [11–13]. Szwedowicz et al. [13] investigated numerically and experimentally the performance of a thin-walled damper mounted under the platform of two rotating free-standing high-pressure turbine blades. Characterization of friction contact of nonspherical contact geometries obeying the Coulomb friction law with constant friction coefficient and constant normal load is proposed by [14]. Firrone et al. [15] undertook forced response studies of bladed disc under platform damper considering both static and dynamic displacements.

This paper presents a methodology which combines three essential phases of friction damping optimization in turbomachinery. The first phase is to develop an analytical model of the blade damper system. The second phase is experimentation and model tuning necessary for response studies; while the third phase is evaluating damper performance and arriving at optimal damper design point. Figure 1 gives the flow chart of complete design methodology. In the present work, we are limiting our discussions to static bench test and using this data along with results from analytical studies, studying the damper performance, and then arriving at an optimal damper design.

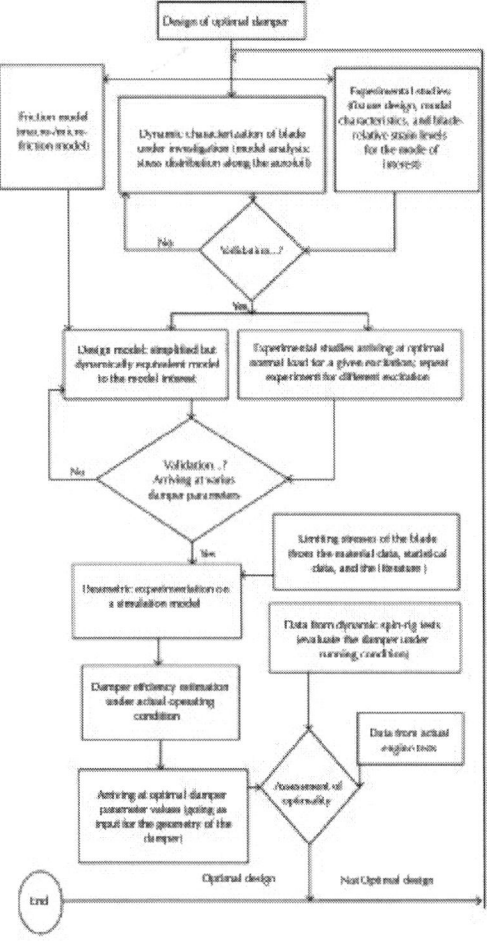

Figure 1: The flow chart indicating the damper design methodology.

DEVELOPMENT OF A MATHEMATICAL MODEL

The development of a mathematical model for a blade damper involves following steps:

- Estimation of eigenvalues and establishment of stress distribution for a specific mode of interest;

- Development of a friction model;
- Conducting response studies using reduced model.

Eigenvalue Estimation and Response Studies

The objective of this phase is to obtain modal characteristics of the blade and stress distribution along the aerofoil for a particular mode of interest along with estimation of nondimensional parameters. Figure 2presents a finite element 3D model of a turbine blade, with boundary condition in the form of root fixation. The blade material is transversally anisotropic with minimal rigidity axis directed along the blade radius.

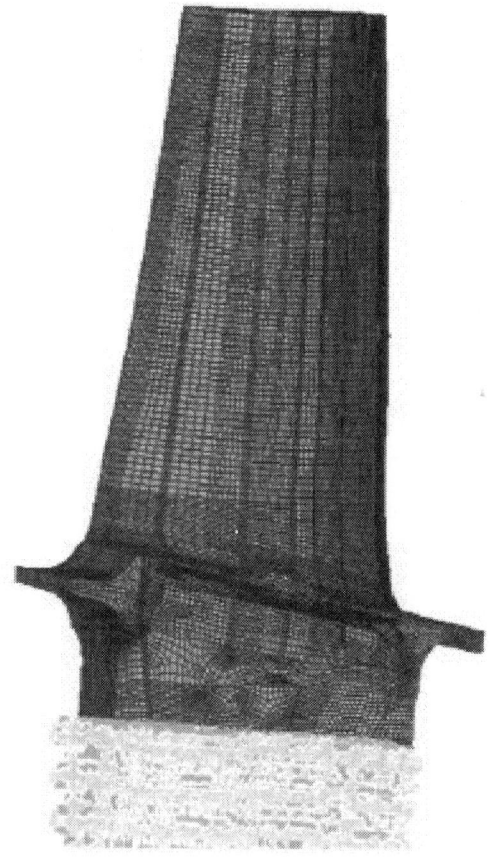

Figure 2: Finite element model of the blade.

Analysis is performed for natural frequencies, distribution of stresses, deformations, and displacements for first six modes. Table 1 gives the various frequencies estimated for various conditions; all the frequencies are normalized with respect to first flexural mode of the blade.

Table 1: Estimated eigenvalues for various conditions normalized with 1st natural frequency at room temperature

Mode number and frequency (Hz)				Simulated condition/constraints
1	2	3	4	
f1	1.74f1	2.26f1	3.06f1	Nonrotating blade at room temperature
1.008f1	1.755f1	2.27f1	3.08f1	Rotating at 50% RPM at room temperature
1.056f1	1.79f1	2.30f1	3.16f1	Rotating at 100% RPM at room temperature
0.95f1	1.61f1	2.06f1	2.85f1	Rotating at 100% RPM at 800°C average blade temperature

The optimization study is undertaken for the first flexure mode of the blade alone; therefore, we are limiting our discussions to this only. Figures 3 and 4 show the 1st flexural mode of the blade and corresponding stress distribution, respectively. The values corroborate well with the experimental results given in Table 2. The maximum stress is located near the fillet area close to the first lobe of the fir tree.

Table 2: Comparison of estimated eigenvalues with experimentally measured values

Mode number	Estimated values	Experimental values
1	f1	1.01f1
2	1.74f1	1.76f1
3	2.26f1	2.27f1
4	3.06f1	3.12f1

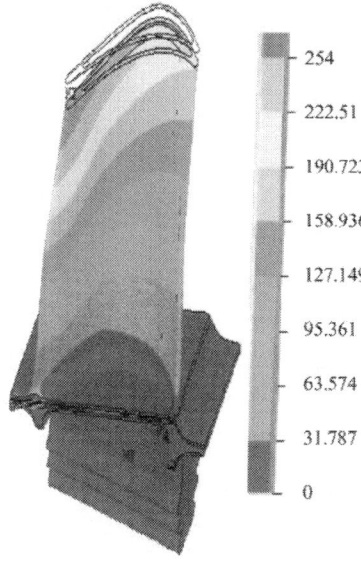

Figure 3: Mode shape for the 1st flexure mode of the blade.

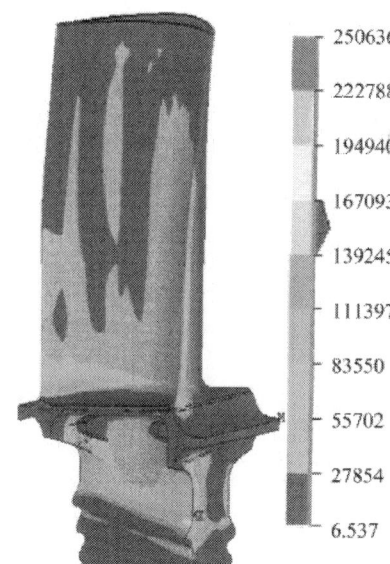

Figure 4: Distribution of equivalent blade stresses at the 1st vibration mode.

With the input drawn from above data, the following coefficients are estimated in order to establish the damper design parameters. These are

- All the natural frequencies normalized with respect to 1st flexural mode of the blade;

- Ratio of maximum stresses in the attachment to stresses in aero foil root $k_{\sigma\sigma} = 1.86$;

- Ratio of maximum total displacement to stresses in aero foil $k_{u\sigma} = 2.12e^{-3}$ mm³/N;

- Ratio of displacement (circumferential direction) at contact location of damping insert to maximum displacement of blade aero foil $k_{uu} = 0.07$;

- Ratio of tip displacement to load amplitude $k_{up} = 0.026$.

Further results from the 3D model calculations were used for identifying strain gauge locations for experimental studies.

Friction Model

The friction interface between the blade root and the damper insert under investigation has a rectangular shape. It is thus appropriate to use a friction interface model that has a rectangular contact surface, such as the model by [5, 9], further the normal load distribution is assumed to be constant over the interface. The model that is used in this paper, shown in Figure 5, is derived from both Menq et al's [5] and the Csaba's one-bar model [9]. The normal load on the damper is assumed as uniform. Further displacement and force is represented in terms of slip length, this feature is derived from Csaba's work [8].

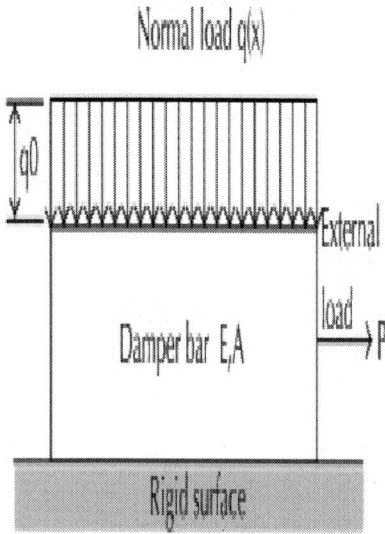

Figure 5: Microslip model for friction interface.

The interface is considered as a rectangular cross-section and modeled as a bar pressed against a rigid surface with a normal load q(x) and the force of P is applied at the end of the bar. The bar has a modulus of elasticity E and area of cross-section A.L, is the length of the bar. The coefficient of friction μ is assumed as constant across the interface and independent of motion of the bar. The normal load across the interface is assumed uniform and constant.

The assumed function of normal load is q(x) =q_0, and the total normal load is found by integrating the normal load function over the length of the bar:

$$N = \int_0^L q(x)dx.$$

(1)

In the present work, the derivation of displacement function is not presented, but relevant equations as and when required are given.

Governing Equation

It is assumed that the bar deforms elastically and the friction is governed by Coulomb's law at each contact point. By applying the load P at the one end of the bar, the onset of slip takes place. Further by increase in amplitude of "P" propagates of slip zone until a gross level slip is achieved. In order to formulate the equation, this bar length is divided into two parts, sliding and not sliding, as shown in Figure 6. The sliding length, that is, the slip zone length is given by "δ" and its maximum value is given by δ_a, corresponding to max force P_a The friction force is defined by F, and its direction depends on the direction of resultant strain. The displacement at the end of bar is defined as "u" The procedure for derivation of displacement function and the force function is followed from [8].

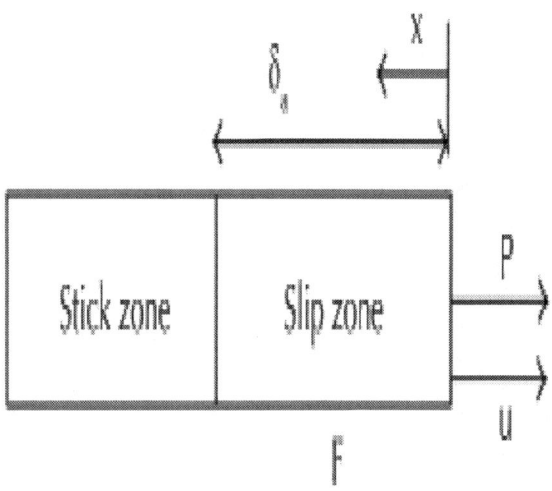

Figure 6: Slip and Stick zones, when the bar is subjected to initial loading.

The force is considered as monotonic between its peak values, with amplitude P_a, then the analysis may be divided into three parts with the bar initially at rest [8]:

P is increased from 0 to P_a,

P is decreased from P_a to $-P_a$,

P is increased from $-P_a$ to P_a,

Formulation of Initial Loading Relations

The Coulomb's friction law says

$$F = \mu q(x).$$

(2)

The various forces acting on a small element of length dx of the sliding zone are given in Figure 7. The direction of "F" depends on whether force "P" is increasing or decreasing with respect to "x"

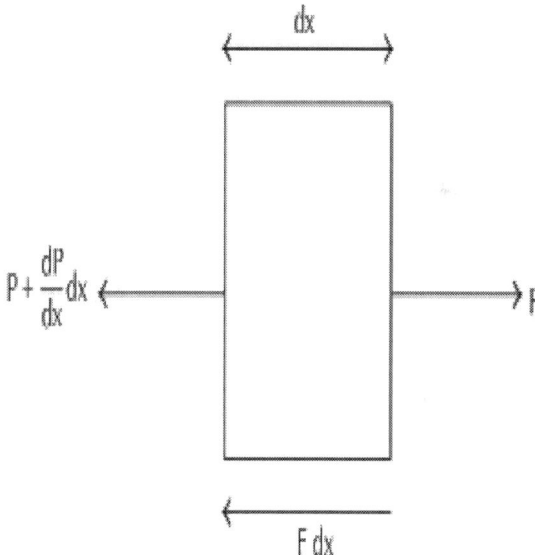

Figure 7: Forces acting on a small element.

Equilibrium of forces yields

$$Fdx + P + \frac{dP}{dx}dx = P.$$

(3)

The strain in bar is

$$\varepsilon = -\frac{du}{dx}.$$

(4)

The tensile force in the bar is defined as

$$P = -EA\frac{du}{dx}\bigg|_{x=0}.$$

(5)

Substituting (5) in (3) we get

$$Fdx - EA\frac{d^2u}{d^2x} = 0,$$

(6)

Where du/dx is the strain in the bar. The displacement "u" is found by double integrating over the slip zone length using the following boundary conditions [8]:

$$x = 0,$$
$$P = -EA\frac{du}{dx}, \qquad x = \delta,\ u = 0.$$

(7)

Therefore, the displacement as a function of force is

$$u(x, P) = P\frac{(x - \delta)}{EA} + \mu q_0 \frac{(\delta^2 - x^2)}{2EA}.$$

(8)

The force at the bar end can be expressed as a function of the slip length. This is done by integrating the friction force over the slip length [8]:

$$P(\delta) = \int_0^\delta F_x dx = \mu q_0 \delta.$$

(9)

The displacement in terms of slip length is given by substituting (9) in (8):

$$u(\delta, x) = \mu q_0 \delta \frac{(x - \delta)}{EA} + \mu q_0 \frac{(\delta^2 - x^2)}{2EA},$$

$$u(\delta, x) = \mu q_0 \frac{(\delta - x)^2}{2EA}.$$

(10)

After having the force and displacement as functions of slip length. Applying "P_{amp}" will give amplitude slip length "$_a$". The force amplitude function is defined by

$$P(\delta_a) = \mu q_0 \delta_a.$$

(11)

And the displacement function is given as

$$u(\delta_a) = \frac{\mu q_0 \delta_a^2}{2EA}.$$

(12)

Load Decreasing Relations

The section of the bar that was slipping is stretched after an initial loading of $P = P_{amp}$. This is given as broken line in Figure 8. The bar is divided into three zones shown in Figure 8, as given in [8]. Zone "Z," where tension is changing in to compression will extend, as the force is decreasing from P_{amp} to $-P_{amp}$. The length of the compression zone length is denoted as $_d$. Zone Z will increase, and zone Y will decrease as P is decreasing, this will continue till $P = -P_{amp}$. Zone Y is then eliminated, and $_d$ equals $_a$.

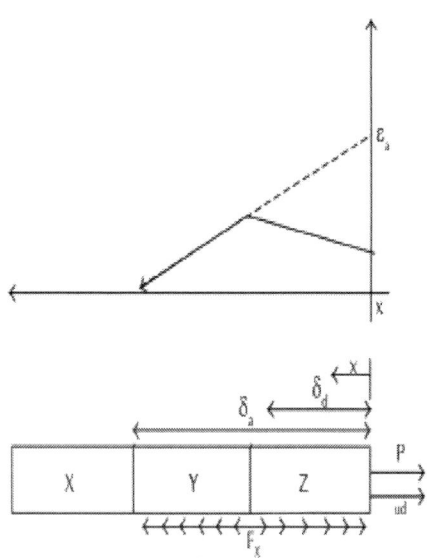

Figure 8: Force, displacement, and plot of strain function, when force is decreasing; zone X is totally struck and has zero strain, zone Y is struck and stretched; zone Z is slipping and compressed.

The friction force direction depends on the sign of dP on that part. The various forces acting on a small element of length dx of zone "Z" is given in Figure 9. The differential equation for this is

$$Fdx + EA\frac{d^2u}{d^2x} = 0.$$

(13)

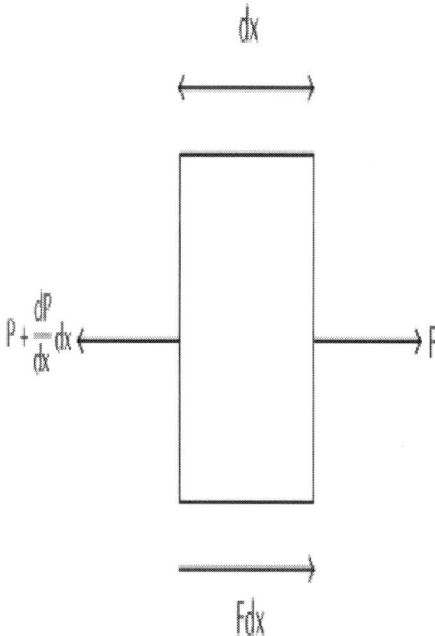

Figure 9: Forces acting on a small element in zone Z, when force is decreasing.

The boundary conditions are found by equilibrium of forces at the bar end, and with a condition that u_a, and u_d must be of the same value, where zone Y and Z are connected. These boundary conditions are, as given in [8]:

$$x = 0,$$
$$P = -EA\frac{du_d}{dx}, \qquad x = \delta_d, \ u_d(\delta_a, \delta_d) = u(\delta_a, \delta_d).$$

$$(14)$$

Solving (13) by substituting (14) yields u_d. The displacement of the bar end will be a function of δ_a and δ_d. The displacement function is given as

$$u_d(\delta_a, \delta_d) = \frac{\mu q_0 \left(\delta_a^2 - \delta_d^2\right)}{2EA}.$$

$$(15)$$

The slip length δ_d is found by equilibrium of forces for the compressed region Z. By equilibrium of forces as shown in Figure 10 and solving one gets the force in terms of slip length. This is given by

$$P_d(\delta_a, \delta_d) = \mu q_0(\delta_a - 2\delta_d),$$

$$(16)$$

Where δ_d is the slip length, when load is decreasing from P_{amp} to $-P_{amp}$.

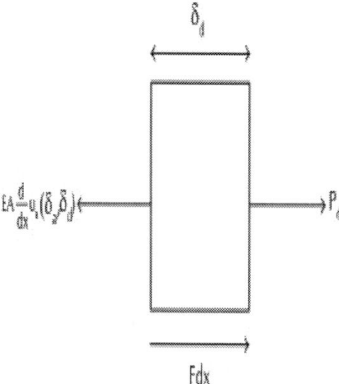

Figure 10: Forces acting on a small element in zone Z, when force is decreasing.

Reloading Relations

This is exactly opposite situation to the one in previous section. Zone Z, where tension will be seen in place of compression, will increase and extend as the force is increasing from $-P_{amp}$ to P_{amp}. The strain in the bar when force P is increasing is shown as the unbroken line in Figure 11, as given in [8].

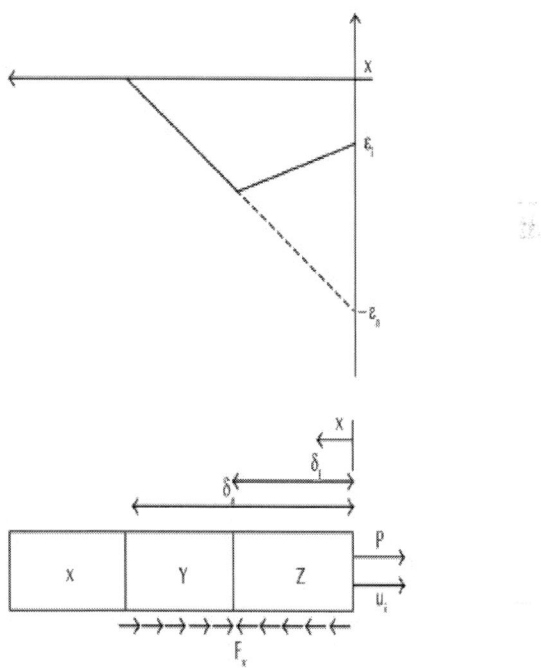

Figure 11: Force, displacement, and plot of strain function, when force is decreasing; zone X is totally struck and has zero strain, zone Y is struck and compressed; zone Z is slipping and stretched.

The length of the stretched zone is denoted by the slip length δ_i. Zone Z will increase, and zone Y will decrease as P is increases. This will continue till $P = P_{amp}$. Zone Y is then eliminated, and δ_i equals δ_a. The differential equation in this case is the same as when the bar is initially loaded. The only difference is the boundary conditions. Here, the boundary conditions are, as given by [8]:

$$Fdx - EA\frac{d^2u}{d^2x} = 0,$$

(17)

$$x = 0,$$
$$P = -EA\frac{du_i}{dx}, \qquad x = \delta_i, \; u_i(\delta_a, \delta_i) = u(\delta_a, \delta_i).$$

(18)

Solving (17) with (18) yields u_i at the bar end as a function of $_a$ and $_i$.

$$u_i(\delta_i, \delta_a) = \frac{\mu q_0\left(\delta_i^2 - \delta_a^2\right)}{2EA},$$
$$P_d(\delta_i, \delta_a) = \mu q_0(2\delta_i - \delta_a),$$

(19)

Where $_i$ is slip length, when the load is increasing from $-P_{amp}$ to P_{amp}. The slip length $_i$ is calculated by equilibrium of forces for the stretched region Y.

Once the relations of both force and displacement at the end of the bar for both unloading and reloading is established in terms of slip length based on initial loading relationships, the hysteretic curve can be generated. Figure 12 is the time domain representation of input force and Figure 13 is simulated displacement at the end of the bar. Using above parametric functions, the hysteretic curve is established and shown in Figure 14. The curve is built up with a starting part (portion OA) for initial loading, the lower curve (portion AB), for unloading $P_d(_d)$ and $u_d(_d)$, and an upper curve (portion BA) for reloading $P_i(_i)$ and $u_i(_i)$. Once the hysteretic curve is established, the damping energy is calculated. The work done per cycle is

$$W = \oint u \, dP.$$

(20)

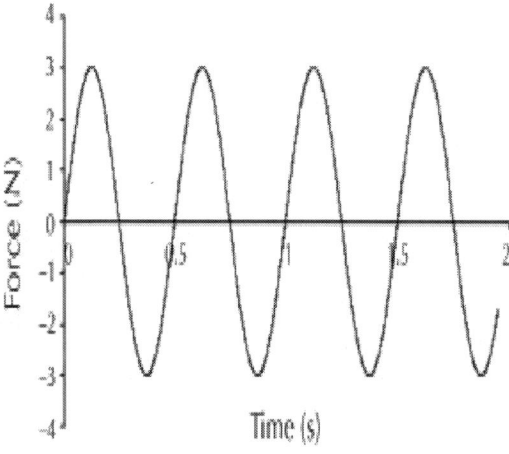

Figure 12: The applied force at the end of the bar.

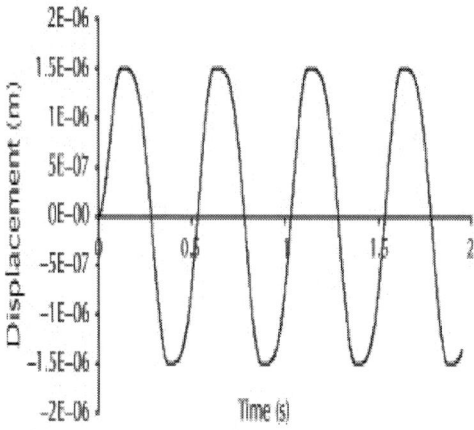

Figure 13: The simulated displacement at the end of the bar.

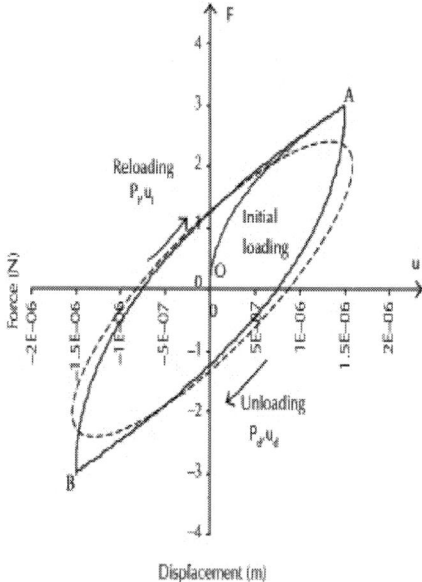

Figure 14: Hysteretic curve drawn between applied force and the simulated displacement at the end of the bar.

Linearization

The objective here is to transform the nonlinear properties of the friction damper into equivalent linear damping and stiffness. The hysteresis loop made by the force versus displacement function is thereby replaced with an equivalent elliptic loop. The linearized properties of the damper will then be used for forced response analysis. Linearization of the friction model is done by using Lazan's method [9].

The work done by equivalent viscous damping is given by

$$D = \pi \omega C_{eq} u_{amp}^2,$$

(21)

By equating (20) and (21), we get

$$C_{eq} = \frac{W}{\pi \omega u_{amp}^2}.$$

(22)

Similarly the equivalent stiffness given by Lazan's method is

$$K_{eq} = \sqrt{\left(\frac{P_{amp}}{u_{amp}}\right)^2 - \left(\omega C_{eq}\right)^2}.$$

(23)

Reduced Model

The dynamically equivalent reduced blade model is represented using a two-degree freedom system simulating 1st flexure mode of blade under investigation, similar to that of given by [8], but additionally the material damping and the aerial damping are represented as a viscous damper. The reason for choosing a two-DOF system for simulations is that it is simple but still illustrates the effect of the damper. At first the response for blade model without damper insert is determined, and correlation is ensured for the first mode between design model and actual blade model. The calculations are performed for various excitation frequency and amplitude of excitation.

The 2-DOF system, shown in Figure 15, is built considering a mass less beam with two concentrated masses simulating aero foil and the root portions of the blade and damper. The aerial and material damping is considered as a viscous damping and attached to the aerofoil portion of the blade. The friction damper is described by an equivalent viscous damper and spring, $C_{eq}(\delta_a)$ and $K_{eq}(\delta_a)$, respectively. Assuming harmonic force of excitation applied at the aerofoil portion of the blade. The displacements where the friction damper is attached and force applied are x_1 and x_2, respectively.

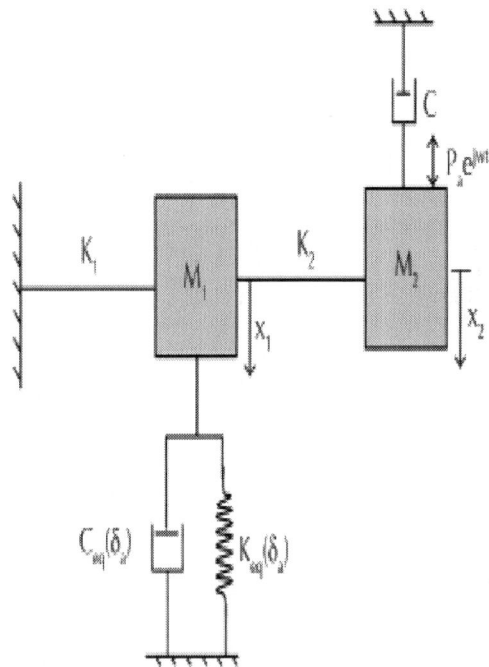

Figure 15: The equivalent blade model, representing two-dof system with friction damper.

As the model is approximate, for obtaining the quantitative estimations, the model is identified using correction factors based on the statistical results obtained from testing the turbine blade and the modal analysis of a 3D model of the blade. Some of the parameters which are used for correcting the model are as follows.

The value of 1st natural frequency of the blade without a damper insert (f_1) has provided the values of mass and stiffness of equivalent model.

The damping in the turbine blade without damper insert is in the range of $\zeta \approx 1\%$ as per the statistics provided by experiments and used as a viscous drag in the model.

The value of the excitation force is chosen such that the amplitude of varying stresses in the blade is in the range of 50–100MP$_a$, this value is estimated from the statistical data available for a typical class of turbine blades.

The friction coefficient is taken in the range of 0.2– 0.4 considering the operating environment of the damper inserts.

The equation of motion for the system is

$$M_1 \ddot{x}_1(t) + C_{eq}(\delta_a)\dot{x}_1(t) + \left(K_1 + K_2 + K_{eq}(\delta_a)\right)x_1(t)$$
$$-K_2 x_2(t) = 0,$$
$$M_2 \ddot{x}_2(t) + C\dot{x}_2(t) + K_2 x_2(t) - K_2 x_1(t) = P_a e^{j\omega t}.$$

$$(24)$$

Assuming harmonic motion:

$$x_1(t) = X_1 e^{j\omega t}, \qquad x_2(t)X_2 e^{j\omega t}.$$

$$(25)$$

And defining the complex stiffness

$$K(\delta_a) = K_{eq}(\delta_a) + j\omega C_{eq}(\delta_a).$$

$$(26)$$

By substituting (26) and (25) in (24) yields algebraic equations:

$$-M_1\omega^2X_1 + (K(\delta_a) + K_1 + K_2)X_1 - K_2X_2 = 0, \tag{27}$$

$$-M_2\omega^2X_2 + C\omega X_2 + K_2X_2 - K_2X_1 = P_a. \tag{28}$$

From (28), $X_2 = (P_a + K_2X_1) / (K_2 + {}_jC - M_2{}^2)$. Substituting X_2 in (27) we get

$$X_1 = \frac{P_aK_2}{(K_2 + jC\omega - M_2\omega^2)((K_1 + K_2 + K(\delta_a) - M_1\omega^2) - K_2^2)},$$

$$X_1 = \frac{P_a}{(K_2 + jC\omega - M_2\omega^2)} + \frac{P_aK_2^2}{(K_2 + jC\omega - M_2\omega^2)((K_1 + K_2 + K(\delta_a) - M_1\omega^2) - K_2^2)}. \tag{29}$$

The response of the system is studied keeping the excitation force amplitude constant, and varying the normal load on the damper.

The solution procedure adopted in the frequency-domain is based on finding the response amplitudes iteratively. The excitation level is selected in such a way that the maximum stress in the blade is in the range of $100\,MP_a$. The starting point being the response levels of the underlying linear system. The behavior of the friction dampers is analyzed at a given relative response amplitude between the damper connection points and the individual dampers. The individual dampers are represented as equivalent complex stiffness, representing both restoring and energy dissipation characteristics as described above. The equivalent complex stiffness is then added to the otherwise linear system, and the response level of the modified system is calculated again. The procedure is repeated till convergence is achieved. The error between successive iterations should be below certain value. In the present case, it is defined as $1E10^{-5}$.

The response levels obtained at current frequency are used as initial guesses for the next frequency increment. Figure 16 shows the normalized responses verses the frequency ratio, under a given excitation for various normalized friction forces. When the friction force is increased the response levels will decrease and attain minimum at particular condition. Further increase in friction force will increase the response amplitude. Figure 17 is the normalized stress versus normalized friction force, and Figure 18 represents plot for damping coefficient verses normalized friction force. The damper design curve/performance curve given in Figure 19, is drawn between normalized response verses normalized excitation, the entire curve can be divided in to three zones, Zone 1 is a completely struck condition. This happens at very low excitation levels or at very high normal loads. In this zone, the system behaves as a linear system. Zone 2 is a slip condition at either low normal loads or at higher excitation level. Here again the system behaves as a linear system. Zone 3 is a slip-stick zone, where the system experiences both conditions in a given cycle, and the system is highly nonlinear in nature.

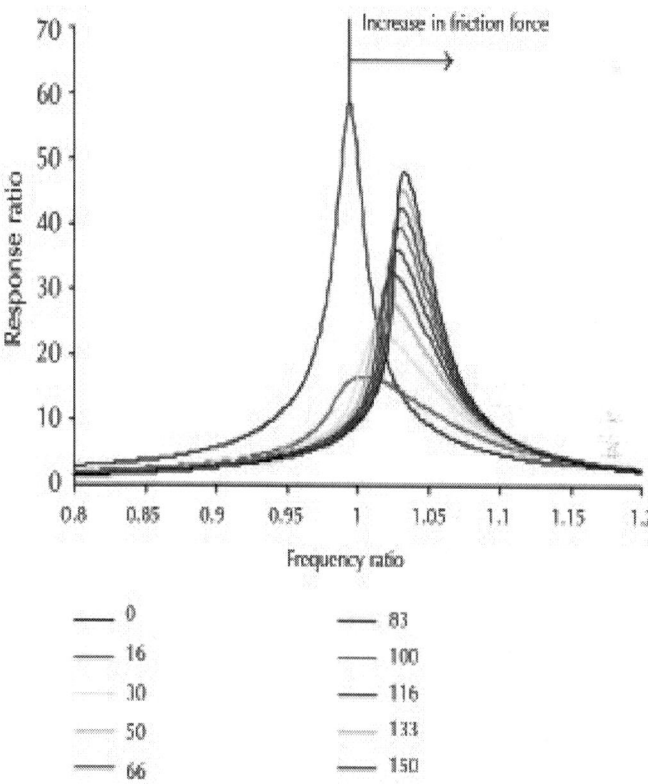

Figure 16: Response ratio versus frequency ratio for various values of normalized friction force.

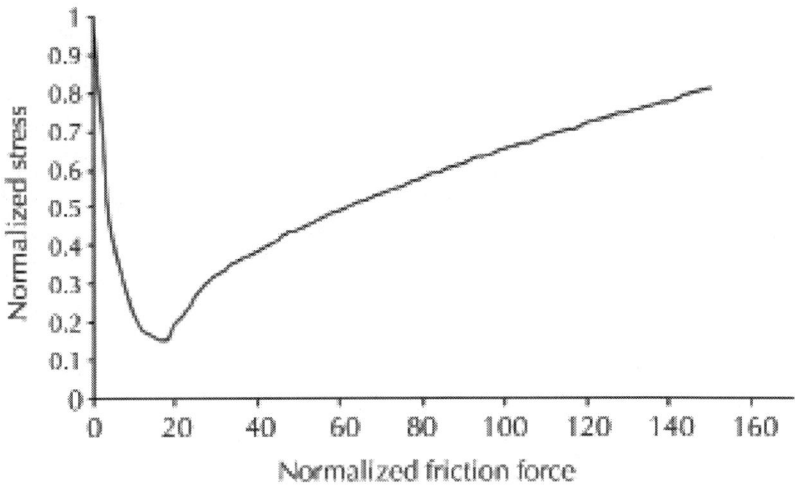

Figure 17: Normalised stress versus normalised friction force.

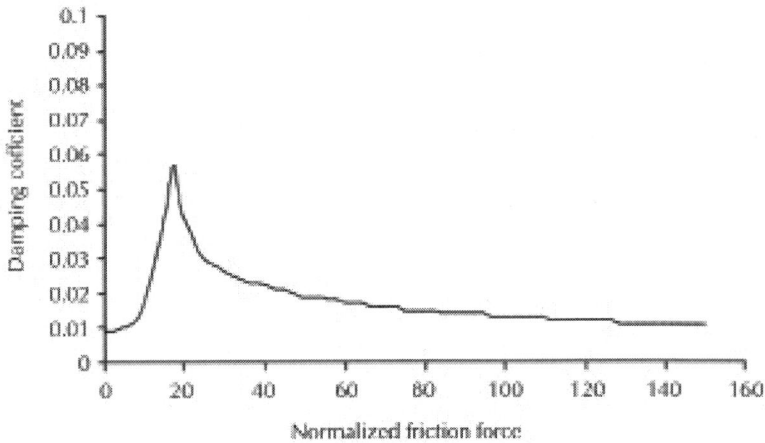

Figure 18: Damping coefficient versus normalized friction force.

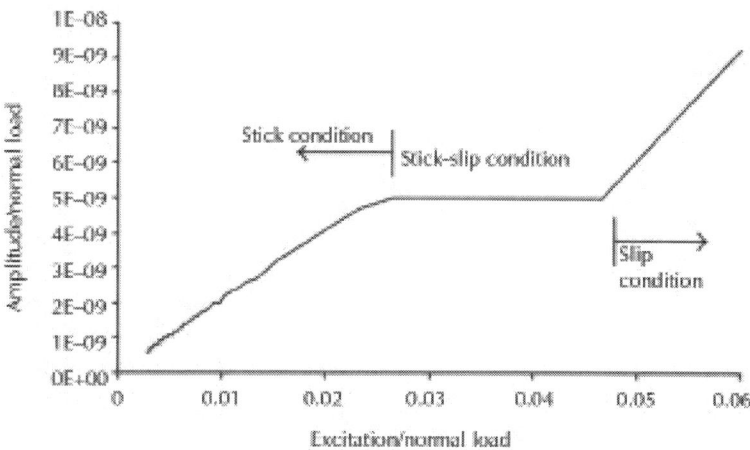

Figure 19: Damper design curve plotted between normalized response and the normalised force.

Figure 20 is the damper performance curve generated for various viscous damping levels. From this figure, it is clear that the damper performance curves depends on the damper properties rather than the normal load and excitation level, which means, for a given damper, one damper performance curve can be generated and that can be used for optimal selection of the damper. These observations are in line with observations made by [2, 3]. The advantage with this representation is that it provides a design point, which is independent of both excitation and viscous damping levels in the engine. The design point is selected corresponding to maximum allowable blade response and consequently, the maximum excitation that can be sustained by the present blade. If actual excitation exceeds this maximum value, then the entire blade must be redesigned since friction damping cannot keep the blade response below allowable limits. Using this approach provides a friction damper that is optimal. The stresses in the blade will be acceptable for as large an excitation as possible. Thus, the optimal design is independent of the excitation and is insensitive to variations in viscous damping.

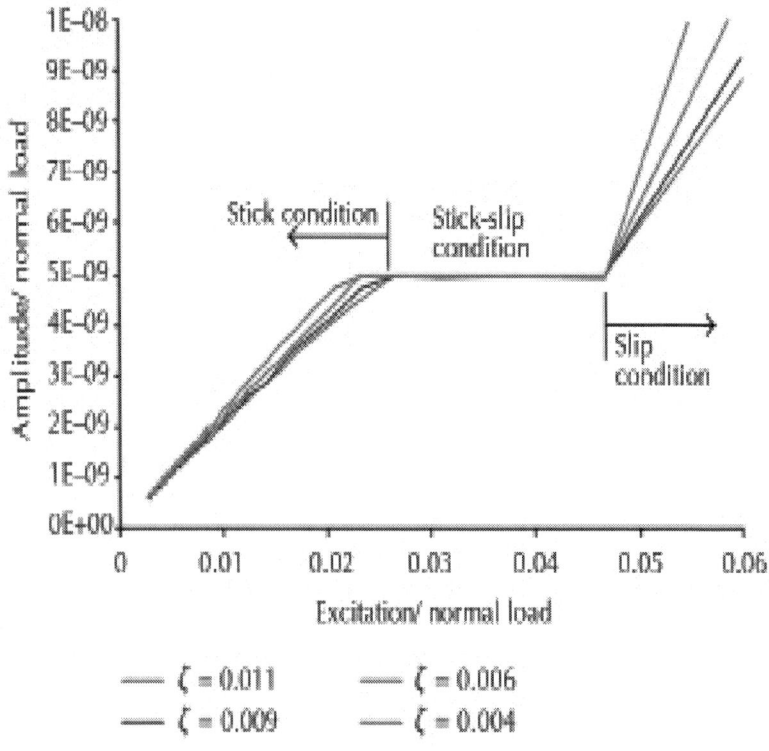

Figure 20: Damper design curve generated under various viscous damping levels.

EXPERIMENTAL STUDIES

The laboratory experiments were carried out in two phases [16]. The first phase establishes the dynamic characteristics of the blade. This includes estimation of natural frequencies, mode shapes, the logarithmic decrement/damping factors, and the relative stress distribution along the aero foil for various conditions. The tests are performed for the free-free condition and the clamped condition. In the second phase, the friction damper is characterized. In the present paper, only the relevant and important results are discussed

Free-Free Condition

In this case, the blade is suspended using a thread and excited using an instrumented impact hammer. The damping estimation by this technique gives only the material damping of the blade (aerial damping due to oscillating blade can be neglected as it is proportionately very small). Knowing this material damping is a must in order to understand the amount of damping blades will experience due to blade fixity in clamped condition.

Clamped Condition

For this a customized fixture is made in order to accommodate the blade under investigation along with the dummy blades on either side with under platform damper inserts. An electrodynamic shaker is used to excite the blade assembly that is driven by the signal generator through the amplifier in the frequency range of interest. The amplitude of excitation is kept constant by using a feedback control loop. The response is measured by blade-mounted strain gages. While placing the strain gauges on the blades, the following aspects are considered:

- Sensor operability is ensured throughout the operating conditions of the blade;
- Instrumentation convenience;
- Informative capability of the strain gauges for the mode of interest.

   The first two conditions are usually satisfied by a strain gauge located on blade suction side near aero foil root. The informative capability of strain gauge is ensured through the 3D FEM analysis and experimental strain survey. One will look for maximum variable stresses for a given oscillation mode. Figure 21 demonstrates the strain gauges pasted on the blade under test. Apart from these strain gages, various other transducers like eddy current probes and accelerometers are used for the rig operation.

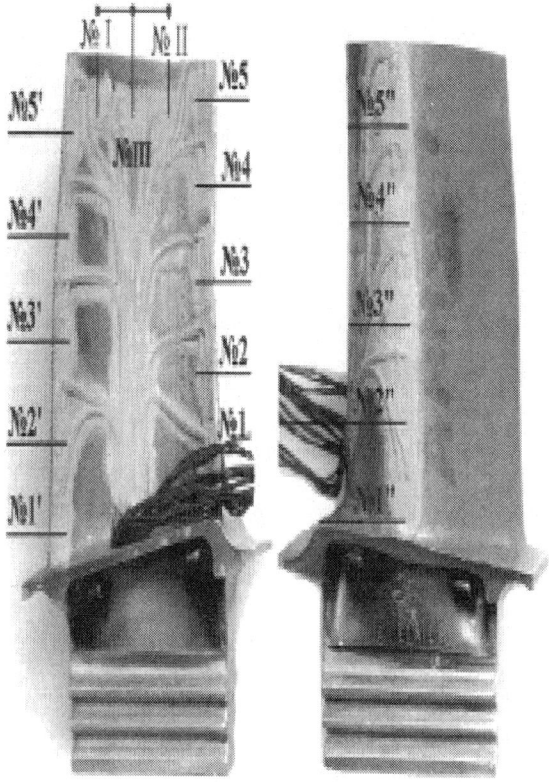

Figure 21: Indicating the strain gauges mounted on both pressure and section side of the blade.

The signals measured from strain gages mounted on blade are used for characterizing the blade. When studying frequency characteristics and the damping of fixed blades, it is necessary to identify frequencies using spectral analysis and then finds its attributes, namely, the one caused by blade's resonating frequencies and other caused by excitation of fixture subcomponents. When fixing blades in spring clamps, blade spectrums are investigated for various attachment designs, for different rigidity, mass, and squeezing efforts. Figure 22 shows the single blade setup under clamped condition, while Figure 23 shows the setup for dynamic characterization of blade damper assembly under clamped condition.

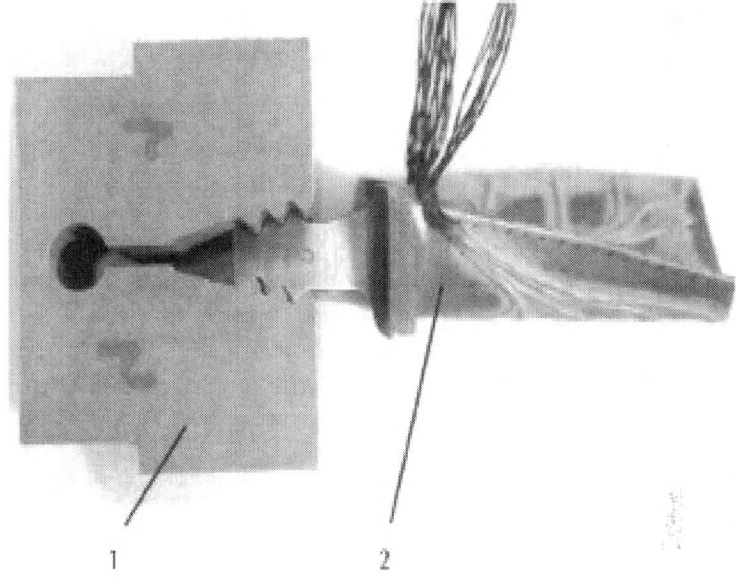

Figure 22: Blade fixation at the root in a spring clamp. (1) Spring clamp and (2) blade under investigation.

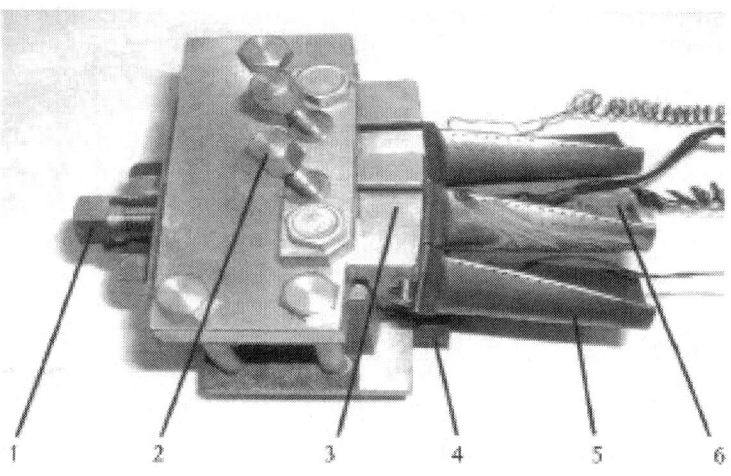

Figure 23: Tooling used for estimation of damper efficiency screw. (2) Load screw, (3) lever, (4) support, (5) blade, and (6) strain gauges.

Table 3 gives measured natural frequency and corresponding damping value for both the fixity conditions. In both of these tests, it is clear that the first flexure mode is in the range of 1.01f1 and1.03f1. The logarithmic decrement is in the range of 0.5 to 1%, which is within the limits of scatter identified from the free-free test on the individual blade. From the test, it is apparent that the energy dissipation due to friction induced due to the clamping arrangement is very small. Certain divergence in frequency values can be attributed to difference in ways of fixation, scatter of blade geometric features within technological tolerance, and difference of real anisotropy of blade material from orthotropic parameters set forth in calculations. Tables 4(a) and 4(b) indicate measured values of relative strain corresponding to first flexure mode of the blade under zero normal load on the damper. From the table, it is very clear that the most loaded area on the blade is near root, and the maximum displacement is measured at leading edge, pressure side of the blade. The relative deformation is

$$\overline{\varepsilon_i} = \frac{\varepsilon_i}{\varepsilon_{max}},$$

(30)

Where i is the strain gauge number, $\overline{\varepsilon_i}$ is the relative deformation in blade aero foil, ε_i is the deformation measured in aero foil by the ith strain gauge, and ε_{max} is the maximum deformation measured by the any strain gauge on blade aero foil.

Table 3: Measured modal parameters for 1st flexural mode at room temperature

Blade fixity	Frequency (Hz)	Log decrement %
Spring clamping of root	1.01f1	0.47%
Three blades fixed in a fixture	1.03f1	0.58%

Table 4: Relative strain at various locations in 1st flexure mode for zero damper load

(a)

Sl number	1	2	3	4	5	6	7	8
Gauge number	1	2	3	4	1'	2'	3'	4'
Relative strain	1	0.64	0.36	—	0.5	0.42	0.34	0.38

(b)

Sl number	9	10	11	12	13	14	15	16
Gauge number	5'	1"	2"	3"	4"	I	II	III
Relative strain	—	0.65	0.61	0.36	—	0.15	0.26	0.2

Characterizing Friction Damper

The test setup for this case remains the same as above except for an additional customized feature to load the damper. The test is executed for various conditions of normal loads at controlled excitation level.

The force level is selected such that the maximum stress experienced by the blade is around 100 MPa. The frequency of excitation is selected around the natural frequency of the blade, and the responses were measured till the steady state achieved. The test is repeated for several damper load conditions. The values of resonance frequencies and its logarithmic decrement at different loads of the damper are given in Table 5. It is observed that the highest logarithmic decrement value is 5.08% at 500 N damper load. The natural frequency of the blade increases with increase in damper load, and to a greater extent in case of first bending mode.

Table 5: Values of natural frequency and damping of 1st flexure mode of the blade under various damper loads

Sl number	Damper load (N)	Frequency in Hz	Percentage
1	0	1.03f1	0.96
2	300	1.095f1	3.25
3	500	1.154f1	5.08
4	1000	1.170f1	2.54
5	2000	1.191f1	2.5

ESTIMATION OF DAMPER EFFICIENCY

With the available experimental and theoretical data, the efficiency of the damper insert is evaluated [17].

The normal load generated by the damper at a given operating speed is

$$N = m\omega^2 r,$$

$$m \times r = 1415\,\mathrm{g} - \mathrm{mm},$$

(31)

Where m is the mass of the damper in "g", and r is the radius at which damper located in "mm." Angular velocity   in rads^{-1}.

Figure 24 is a Campbell plot for a blade. It shows two potential resonance conditions corresponding to first flexure mode, due to 6th- and 9th-order cross-overs, at 12000 rpm and 16000 rpm, respectively.

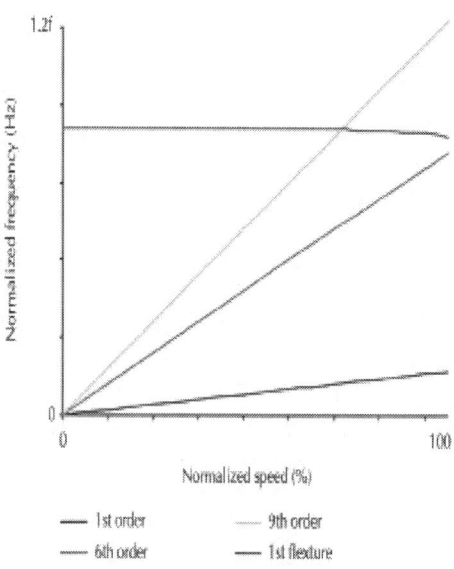

Figure 24: Campbell diagram, indicating 1st flexure mode with various excitation lines.

The normal load corresponding to 9th-order cross-over is $N = 1834$ N.

A typical value of friction coefficient for metal-metal contact under slow slide is $\mu = 0.2$. This value under vibrating contact surfaces is known to reduce the contact and the friction ratios. Contrast to this, under high-frequency relative motion of the bodies, frictional seizure phenomena, due to heating of local zones can occur, leading to increase of friction coefficient. Therefore for calculations, a broad range of possible values for μ ranging from 0.2 to 0.4 is considered. The corresponding friction force F for this will be in the range of 180–734 N.

Using various factors generated from the finite element analysis the range of normalized friction force is identified. The ratio of tip displacement to resonant stresses is $k_u = 2.12e^{-3}$ mm³/N. Considering the resonance stresses in the range of 50–100 Mp_a, the displacement of the blade tip is in the range of 0.11 mm–0.2 mm. Once the blade tip displacement is identified using $k_{uP} = 0.026$, the range of excitation P can be identified. Therefore, excitation P is in the range of 4–8 N.

And the range of normalized friction force is calculated as

$$\left(\overline{F}\right)_{cal} = \frac{F}{P} \text{ is } 22 \text{ to } 174.$$

(32)

Keeping all other parameters constant the $(\overline{F})$ is normalized friction force, proportional to the normal load applied on the damper.

Therefore, $(\overline{F})/N = cont'$.

The optimum range of normalized friction force corresponding to maximum rotational frequency can be found from

$$\left(\overline{F}\right)_{opt} = \frac{N_{opt}}{N}\left(\overline{F}\right)_{cal},$$

(33)

Where N_{opt} is the optimum damper load at which the response of the blade is minimum, that is, 500 N, and N is the normal load on the damper at a given rotational speed.

Therefore, the $(\overline{F})_{opt}$ is in the range of 6 to 47. In this range of friction forces significant reduction in stresses is expected. Further increase in normal load may increase the aero foil stresses.

COMPARISON OF ANALYTICAL AND EXPERIMENTAL FINDINGS

The correlation studies is carried out in two phases, in phase one, the dynamic characteristics of the blade under investigation is carried out, which includes the natural frequency, mode shape, and the damping. Further it is ensured that the blade fixity will have negligible effect on the damping. The relative strain estimated along the aero foil for the first flexure mode is in good correlation with experiments conducted without damper inserts. The damper effectiveness is measured on bench test is in good agreement with the analytical findings. The maximum logarithmic decrement value measured with damper insert is around 5.08%, in case of analytical estimation it is around 5.8%. From the experiments it is found that the optimal normalized friction force is in the range of 6–47, for the blade damper system under investigation, corresponding to the rotational speed of 16000 rpm, where the possibility of 1st flexure mode resonance exists. Similarly in the analytical solution, it is found to be in the range of 7–50, refer to Figure 16. It is clear that the range of normalized friction force for which the response amplitude is below 50% of resonance amplitude without a damper insert. Figure 25 indicates the comparison of both theoretically estimated and the experimentally measured damping coefficient

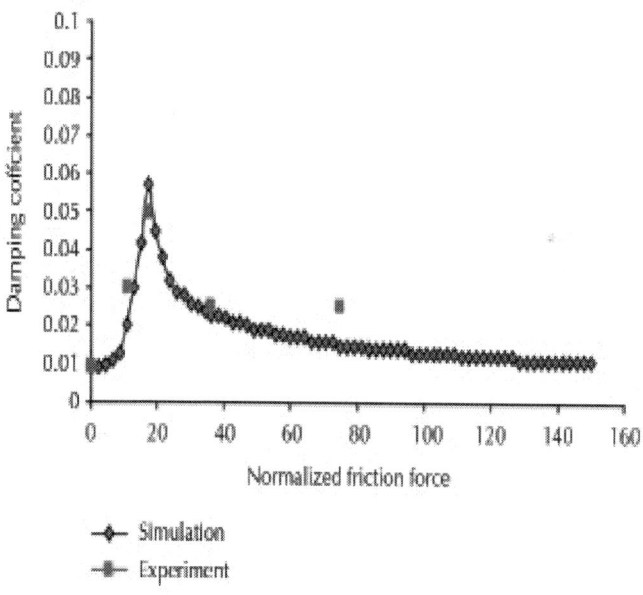

Figure 25: Comparison of damping coefficient value with normalized friction force.

CONCLUSIONS

Minimization of resonant stresses in turbine blades is a major concern in turbine engine. A detailed procedure is outlined for optimization of damper for turbine blade so that the maximum stress experienced by any blade should be below some designated maximum value. The procedure details the integration of analytical and experimental studies to ensure the correctness of the model developed and the friction damper design. A damper performance curve is established, which provides a design point that is independent of both damping and the excitation levels, these two quantities are very difficult to be determined for the new design of blades. At the same time, the design point optimizes the friction damper for as large an excitation as possible. A unique test fixture is developed for conducting the experiments, having a feature to setup various damper loads.

The predicted natural frequencies, stress distribution, and the displacements determined for the 1st flexure mode of the blade is in good correlation with the experimental results. The damper performance curves are generated for various levels of force and the damping values, which are used to arrive at optimal damper design. This can now be extended to the engine condition, namely, gas loads, and temperatures.

ACKNOWLEDGMENTS

The authors would like to thank the Director of GTRE, for his support throughout this work. And authors also would like to thank the Head of Mechanical Engineering Department, SV University, for his encouragement and guidance and Mr. Sreelal Sreedhar, Additional Director, GTRE, for his valuable inputs. Finally authors thank one and all whomever directly and indirectly involved and helped during this work.

REFERENCES

1.   J. H. Griffin, "A review of friction damping of turbine blade vibration," International Journal of Turbo and Jet Engines, vol. 7,

pp. 297–307, 1990.

2. T. M. Cameron, J. H. Griffin, R. E. Kilb, and T. M. Hoosac, "An integrated approach for friction damper design," The Role of Damping in Vibration and Noise Control, vol. 5, pp. 205–211, ASME Booklet DE, 1987.

3. R. Kilb, J. Griffin, and C. H. Menq , "Evaluation of a turbine blade damper using an integral approach," in Proceedings of the 29th Structures, Structural Dynamics and Materials Conference AIAA/ASME/ASCE/AHS, no. 88-2400, AIAA, Williamsburg, Va, USA, 1998.

4. K. Y. Sanliturk, D. J. Ewins, R. Elliott, and J. S. Green, "Friction damper optimization: simulation of rainbow tests," Journal of Engineering for Gas Turbines and Power, vol. 123, no. 4, pp. 930–939, 2001.

5. C. H. Menq, J. H. Griffin, and J. Bielak, "Influence of a variable normal load on the forced vibration of a frictionally damped structure," Journal of Engineering for Gas Turbines and Power, vol. 108, no. 2, pp. 300–305, 1986.

6. C. H. Menq, J. Bielak, and J. H. Griffin, "The influence of microslip on vibratory response, part I: a new microslip model," Journal of Sound and Vibration, vol. 107, no. 2, pp. 279–293, 1986.

7. C. H. Menq, J. Bielak, and J. H. Griffin, "The influence of microslip on vibratory response, Part II: a comparison with experimental results," Journal of Sound and Vibration, vol. 107, no. 2, pp. 295–307, 1986.

8. G. Csaba, Modelling microslip friction damping and its influence on turbine blade vibrations, Dissertations no. 519, Linköping, Sweden, 1998.

9. G. Csaba, "Forced response analysis in time and frequency domains of a tuned bladed disk with friction dampers," Journal of Sound and Vibration, vol. 214, no. 3, pp. 395–412, 1998.

10. E. Cigeroglu and H. N. Ozguven, "Nonlinear vibration analysis of bladed disks with dry friction dampers," Journal of Sound and Vibration, vol. 295, no. 3–5, pp. 1028–1043, 2006.

11. Schwingshackl C. W., et al., "Measured and estimated friction interface parameters in a nonlinear dynamic analysis," Mechanical Systems and Signal Processing, vol. 28, pp. 574–584, 2012.

12. E. P. Petrov and D. J. Ewins, "Advanced modeling of underplatform friction dampers for analysis of bladed disk vibration," Journal of Turbomachinery, vol. 129, no. 1, pp. 143–150, 2007.

13. J. Szwedowicz, C. Gibert, T. P. Sommer, and R. Kellerer, "Numerical and experimental damping assessment of a thin-walled friction damper in the rotating setup with high pressure turbine blades,"Journal of Engineering for Gas Turbines and Power, vol. 130, no. 1, Article ID 012502, 2008.

14. M. Allara, "A model for the characterization of friction contacts in turbine blades," Journal of Sound and Vibration, vol. 320, no. 3, pp. 527–544, 2009.

15. C. M. Firrone, S. Zucca, and M. M. Gola, "The effect of under platform dampers on the forced response of bladed disks by a coupled static/dynamic harmonic balance method," International Journal of Non-Linear Mechanics, vol. 46, no. 2, 2011.

16. Vibration Engineering Group, "Experimental investigation on turbine blade damper," Internal Report, GTRE, Bangalore, India.

17. Vibration Engineering Group, "Turbine blade damper optimization studies," Internal Document, GTRE, Bangalore, India.

Chapter 5

Fault Detection and Diagnosis for Gas Turbines Based on a Kernelized Information Entropy Model

Weiying Wang[1, 2], Zhiqiang Xu[1, 2], Rui Tang[2], Shuying Li[2], and Wei Wu[3]

[1]College of Power and Energy Engineering, Harbin Engineering University, Harbin 150001, China

[2]Harbin Marine Boiler & Turbine Research Institute, Harbin 150036, China

[3]Harbin Institute of Technology, Harbin 150001, China

ABSTRACT

Gas turbines are considered as one kind of the most important devices in power engineering and have been widely used in power generation, airplanes, and naval ships and also in oil drilling platforms. However, they are monitored without man on duty in the most cases. It is highly desirable to develop techniques and systems to remotely monitor their conditions and analyze their faults. In this work, we introduce a remote system for online condition monitoring and fault diagnosis of gas turbine on offshore oil well drilling platforms based on a kernelized information entropy model. Shannon information entropy is generalized for measuring the uniformity of exhaust temperatures, which reflect the overall states of the gas paths of gas turbine. In addition, we also extend the entropy to compute the information quantity of features in kernel spaces, which help to select the informative features for a certain recognition task. Finally, we introduce the information entropy based decision tree algorithm to extract rules from fault samples. The experiments on some real-world data show the effectiveness of the proposed algorithms.

INTRODUCTION

Gas turbines, mechanical systems operating on a thermodynamic cycle, usually with air as the working fluid, are considered as one kind of the most important devices in power engineering, where the air is compressed, mixed with fuel, and burnt in a combustor, with the generated hot gas expanded through a turbine to generate power, which is used for driving the compressor and for providing the means to overcome external loads. Gas turbines play an increasingly important role in the domains of mechanical drives in the oil and gas sectors, electricity generation in the power sector, and propulsion systems in the aerospace and marine sectors.

Safety and economy are always two fundamentally important factors in designing, producing, and operating gas turbine systems. Once a malfunction occurs to a gas turbine, a serious accident, even disaster, may take place. It was reported that about 25 accidents take place every year due to jet malfunctioning. In 1989, 111 were killed in a plane crash due to an engine fault. Although great progress has been made these years in the area of condition monitoring and fault diagnosis, how to predict and detect malfunctions is still an open problem for the complex systems. In some cases, such as offshore oil well drilling platforms, the main power system is self-monitoring without man on duty. So the reliability and stabilization are of critical importance to these systems. There are hundreds of offshore platforms with gas turbines providing electricity and powers in China. There is an urgent requirement to design and develop online remote monitoring and health management techniques for these systems.

More than two hundred sensors are installed in each gas turbine for monitoring the state of a gas turbine. The data gathered by these sensors reflects the state and trend of the system. If we build a center to monitor two hundred gas turbine systems, we should watch the data coming from more than forty thousand sensors. Obviously, it is infeasible to manually analyze them. Techniques on intelligent data analysis have been employed in gas turbine monitoring and diagnosis. In 2007, Wang et al. designed a conceptual system for remote monitoring and fault diagnosis of gas turbine-based power generation systems [1]. In 2008, Donat et al. discussed the issue of data visualization, data reduction, and ensemble learning for intelligent fault diagnosis in gas turbine engines [2]. In 2009, Li and Nilkitsaranont described a prognostic approach to estimating the remaining useful life of gas turbine engines before their next major overhaul based on a combined regression technique with both linear and quadratic models [3]. In the same year, Bassily et al. proposed a technique, which assessed whether or not the multivariate autocovariance functions of two independently sampled signals coincide, to detect faults in a gas turbine [4]. In 2010, Young et al. presented an offline fault diagnosis method for industrial gas turbines in a steady-state using Bayesian data analysis. The authors employed multiple Bayesian models via model averaging for improving the performance of the resulted system [5]. In 2011, Yu et al. designed a sensor fault diagnosis technique for Micro-Gas Turbine Engine based on wavelet entropy, where wavelet decomposition was utilized to decompose the signal in different scales, and then the instantaneous wavelet energy entropy and instantaneous wavelet singular entropy are computed based on the previous wavelet entropy theory [6].

In recent years, signal processing and data mining techniques are combined to extract knowledge and build models for fault diagnosis. In 2012, Wu et al. studied the issue of bearing fault diagnosis based on multiscale permutation entropy and support vector machine [7]. In 2013, they designed a technique for defecting diagnostics based on multiscale analysis and support vector machines [8]. Nozari et al. presented a model-based robust fault detection and isolation method with a hybrid structure, where time-delay multilayer perceptron models, local linear neurofuzzy models, and linear model tree were used in the system [9]. Sarkar et al. [10] designed symbolic dynamic filtering by optimally partitioning sensor observation, and the objective is to reduce the effects of sensor noise level variation and magnify the system fault signatures. Feature extraction and pattern classification are used for fault detection in aircraft gas turbine engines.

Entropy is a fundamental concept in the domains of information theory and thermodynamics. It was first defined to be a measure of progressing towards thermodynamic equilibrium; then it was introduced in information theory by Shannon [11] as a measure of the amount of information that is missing before reception. This concept gets popular in both domains [12–16]. Now it is widely used in machine learning and data driven modeling [17, 18]. In 2011, a new measurement, called maximal information coefficient, was reported. This function can be used to discover the association between two random variables [19]. However, it cannot be used to compute the relevance between feature sets.

In this work, we will develop techniques to detect abnormality and analyze faults based on a generalized information entropy model. Moreover, we also describe a system for state monitoring of gas turbines on offshore oil well drilling platforms. First we will describe a system developed for remote and online condition monitoring and fault diagnosis of gas turbines installed on oil drilling platforms. As vast amount of historical records is gathered in this system, it is an urgent task to design algorithms for automatically online detecting abnormality of the data and analyze the data to obtain the causes and sources of faults. Due to the complexity of gas turbine systems, we focus on the gas-path subsystem in this work. The function of entropy is employed to measure the uniformity of exhaust temperatures, which is a key factor reflecting the health of the gas path of a gas turbine and also reflecting the performance of the gas turbine. Then we extract features from the healthy and abnormal records. An extended information entropy model is introduced to evaluate the quality of these features for selecting informative attributes. Finally, the selected features are used to build models for automatic fault recognition, where support vector machines [20] and C4.5 are considered. Real-world data are collected to show the effectiveness of the proposed techniques.

The remainder of the work is organized as follows. Section 2 describes the architecture of the remote monitoring and fault diagnosis center for gas turbines installed on the oil drilling platforms. Section 3designs an algorithm for detecting abnormality of the exhaust temperatures. Then we extract features from the exhaust temperature data and select informative ones based on evaluating the information bottlenecks with extend information entropy in Section 4. Support vector machines and C4.5 are introduced for building fault diagnosis models in Section 5. In addition, numerical experiments are also described in this section. Finally, conclusions and future work are given in Section 6.

FRAMEWORK OF REMOTE MONITORING AND FAULT DIAGNOSIS CENTER FOR GAS TURBINE

Gas turbines are widely used as power and electric power sources. The structure of a general gas turbine is presented in Figure 1. This system transforms chemical energy into thermal power, then mechanical energy, and finally electric energy. Gas turbines are usually considered as the hearts of a lot of mechanical systems.

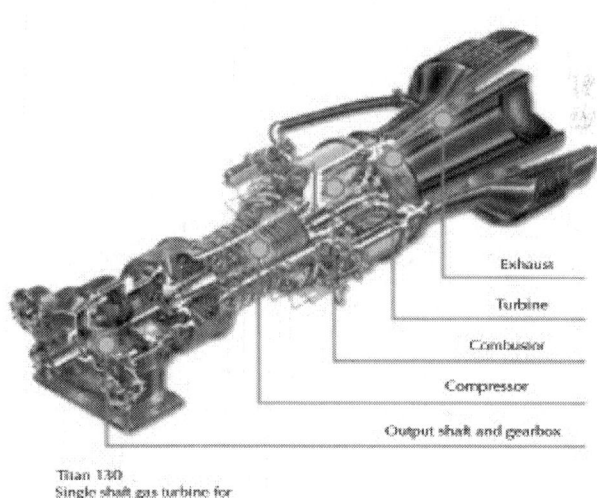

Figure 1: Prototype structure of a gas turbine.

As the offshore oil well drilling platforms are usually unattended, an online and remote state monitoring system is much useful in this area, which can help find abnormality before serious faults occur. However, the sensor data cannot be sent into a center with ground based internet. The data can only be transmitted via telecommunication satellite, which was too expensive in the past. Now this is available.

The system consists of four subsystems: data acquisition and local monitoring subsystem (DALM), data communication subsystem (DAC), data management subsystem (DMS), and intelligent diagnosis system (IDS). The first subsystem gathers the outputs from different sensors and checks whether there is any abnormality in the system. The second one packs the acquired data and transforms them into the monitoring center. Users in the center can also send a message to this subsystem to ask for some special data if abnormality or fault occurs. The data management subsystem stores the historic information and also fault data and fault cases. A data compression algorithm is embedded in the system. As most of the historic data are useless for the final analysis, they will be compressed and removed for saving storage space. Finally, IDS watches the alarm information from different unit assemblies and starts the corresponding module to analyze the related information. This system gives some decision and explains how the decision has been made. The structure of the system is shown in Figure 2.

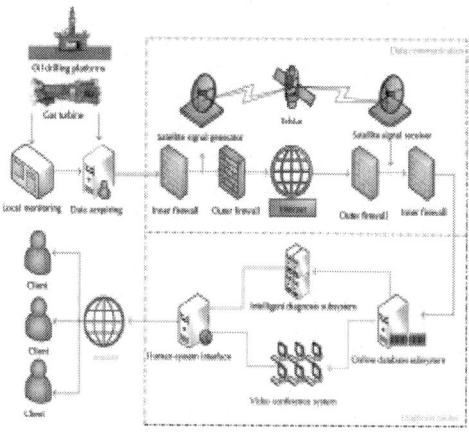

Figure 2: Structure of the remote system of condition monitoring and fault analysis.

One of the webpages of the system is given in Figure 3, where we can see the rose figure of exhaust temperatures, and some statistical parameters varying with time are also presented.

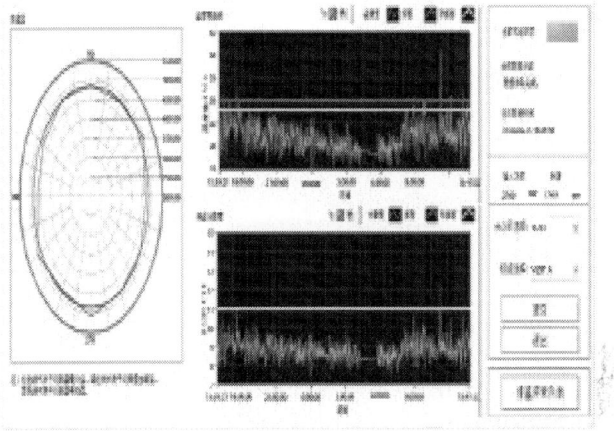

Figure 3: A typical webpage for monitoring of the subsystem.

ABNORMALITY DETECTION IN EXHAUST TEMPERATURES BASED ON INFORMATION ENTROPY

Exhaust temperature is one of the most critical parameters in a gas turbine as excessive turbine temperatures may lead to life reduction or catastrophic failures. In the current generation of machines, temperatures at the combustor discharge are too high for the type of instrumentation available. Exhaust temperature is also used as an indicator of turbine inlet temperature.

As the temperature profile out of a gas turbine is not uniform, a number of probes will help pinpoint disturbances or malfunctions in the gas turbine by highlighting the shifts in the temperature profile. Thus there are usually a set of thermometers fixed on the exhaust. If the system is normally operating, all the thermometers give similar outputs. However, if a fault occurs to some components of the turbine, different temperatures will be observed. The uniformity of exhaust temperatures reflects the state of the system. So we should develop an index to measure the uniformity of the exhaust temperatures. In this work, we consider the entropy function for it is widely used in measuring uniformity of random variables. However, to the best of our knowledge, this function has not been used in this domain.

Assume that there are n thermometers and their outputs are T_i, $i = 1,\ldots,$ respectively. Then we define the uniformity of these outputs as

$$E(T) = -\sum_{i=1}^{n} \frac{T_i}{T} \log_2 \frac{T_i}{T},$$

(1)

where $T=\sum_j T_j$. As $T_i \geq 0$, we define $0 \log 0 = 0$. Obviously, we have $\log_2 n \geq (T) \geq 0$. $E(T) = \log_2 n$ if and only if $T_1 = T_2 = \cdots = T_n$. In this case, all the thermometers produce the same output. So the uniformity of the sensors is maximal. In another extreme case, if $T_1 = T_2 = T_{i-1} = T_{i+1} \cdots = T_n = 0$ and $T_i = T$, then $E(T) = 0$.

It is notable that the value of entropy is independent of the values of thermometers, while it depends on the distribution of the temperatures. The entropy is maximal if all the thermometers output the same values.

Now we show two sets of real exhaust temperatures measured on an oil well drilling platform, where 13 thermometers are fixed. In the first set, the gas turbine starts from a time point and then runs for several minutes; finally the system stops.

Observing the curves in Figure 4, we can see that the 13 thermometers give the almost the same outputs at the beginning. In fact, the outputs are the room temperature in this case, as shown in Figure 6(a). Thus, the entropy reaches the peak value.

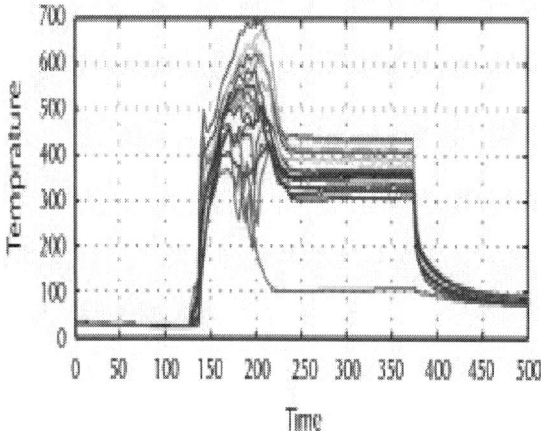

Figure 4: Exhaust temperatures from a set of thermometers.

Some typical samples are presented in Figure 6, where the temperature distributions around the exhaust at time points $t=5,130,250,400$, and 500 are given. Obviously, the distributions at $t = 130, 250$, and 400 are not desirable. It can be derived that some abnormality occurs to the system. The entropy of temperature distribution is given in Figure 5.

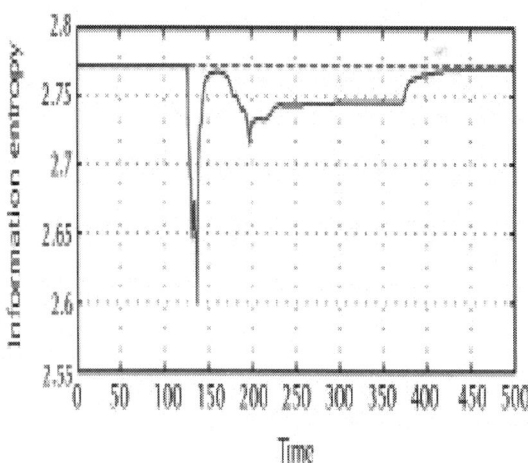

Figure 5: Uniformity of the temperatures (red dash line is the ideal case; blue line is the real case).

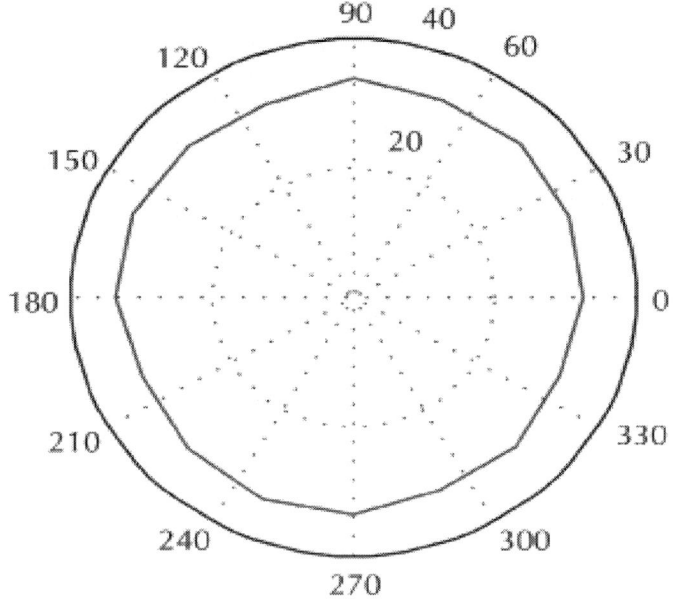

(a)

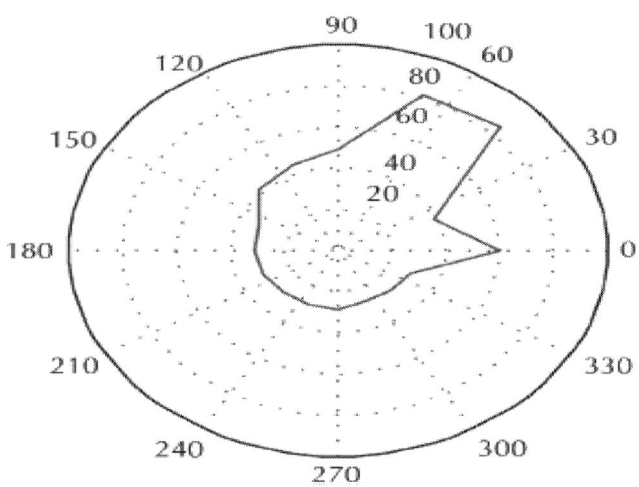

(b)

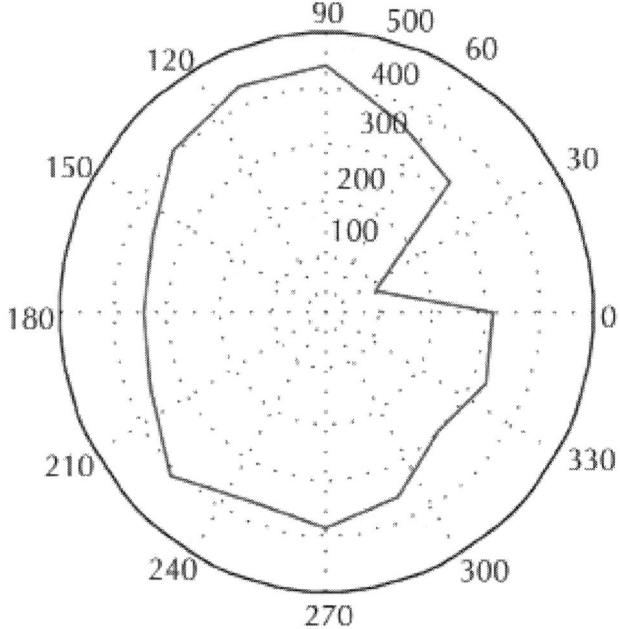

(c)

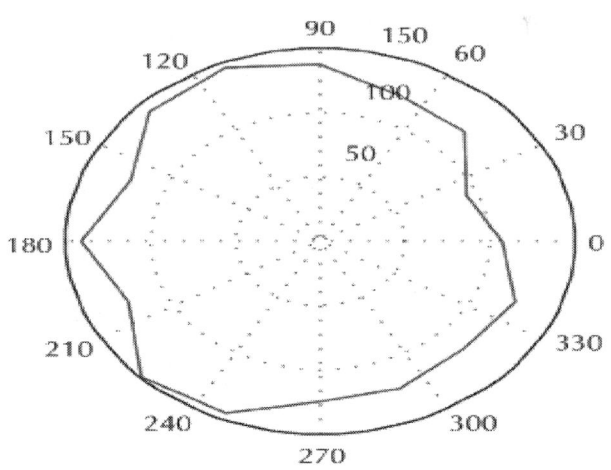

(d)

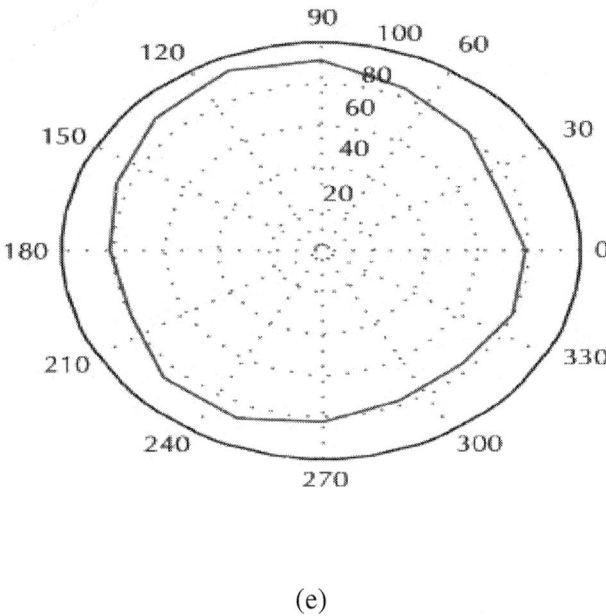

(e)

Figure 6: Samples of temperature distribution in different times.

Another example is also given in Figures 7 to 9. In this example, there is significant difference between the outputs of 13 thermometers even when the gas turbine is not running, just as shown in Figure 9(a). Thus the entropy of temperature distribution is a little lower than the ideal case, as shown in Figure 8. Besides, some representative samples are also given in Figure 9.

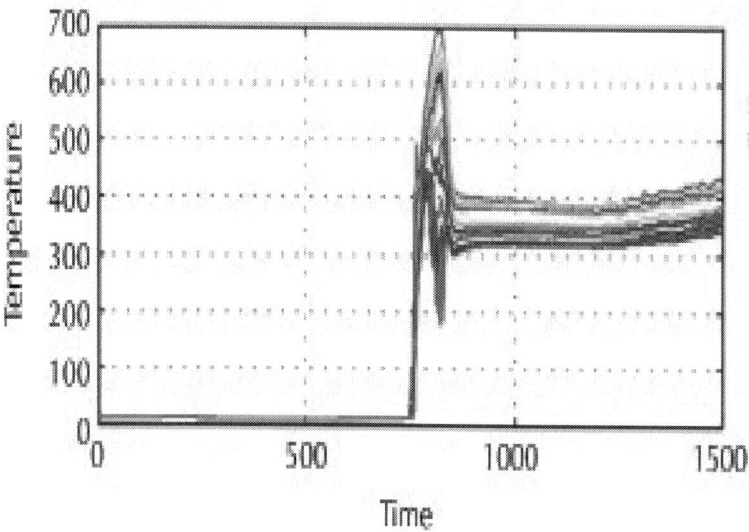

Figure 7: Exhaust temperatures from another set of thermometers.

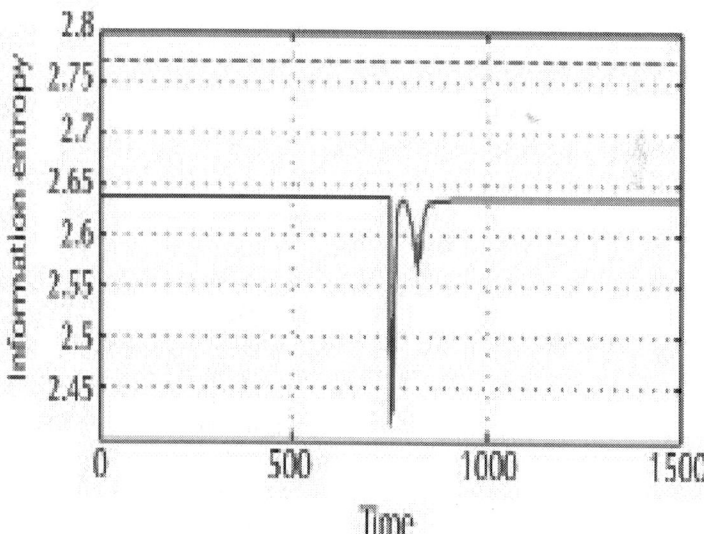

Figure 8: Entropy of the temperature distribution, where the red dash line is the ideal case and the blue one is the real case.

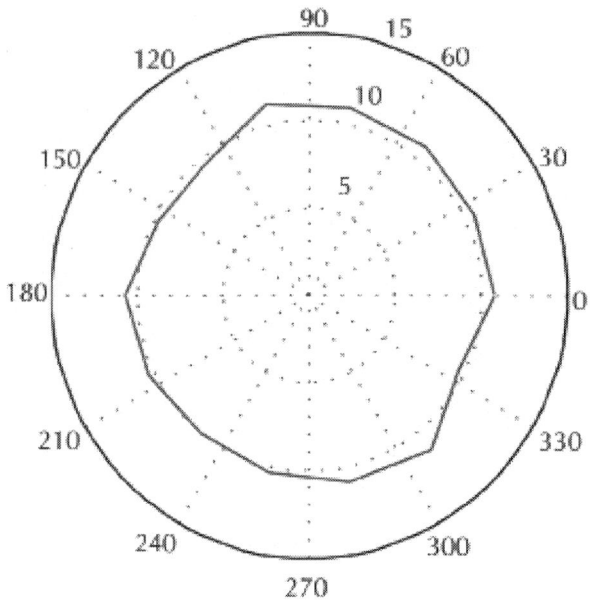

(a)

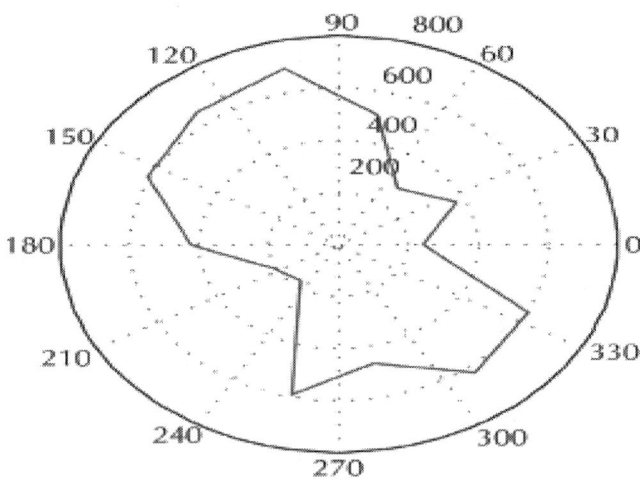

(b)

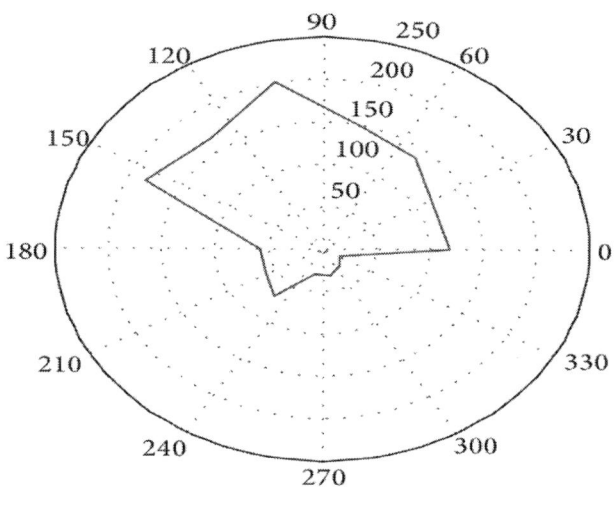

(c)

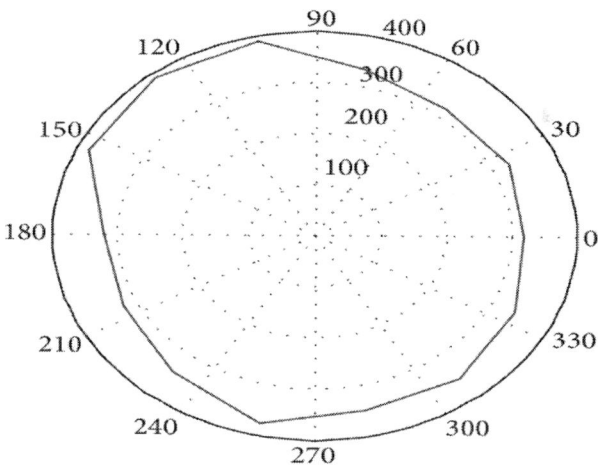

(d)

Figure 9: Samples of temperature distribution in different moments.

Considering the above examples, we can see that the function of entropy is an effective measurement of uniformity. It can be used to reflect the uniformity of exhaust temperatures. If the uniformity is less than a threshold, some faults possibly occur to the gas path of the gas turbine. Thus the entropy function is used as an index of the health of the gas path.

FAULT FEATURE QUALITY EVALUATION WITH GENERALIZED ENTROPY

The above section gives an approach to detecting the abnormality in the exhaust temperature distribution. However, the function of entropy cannot distinguish what kind of faults occurs to the system although it detects abnormality. In order to analyze why the temperature distribution is not uniform, we should develop some algorithms to recognize the fault.

Before training an intelligent model, we should construct some features and select the most informative subsets to represent different faults. In this section, we will discuss this issue.

Intuitively, we know that the temperatures of all thermometers reflect the state of the system. Besides, the temperature difference between neighboring thermometers also indicates the source of faults, which are considered as space neighboring information. Moreover, we know the temperature change of a thermometer necessarily gives hints to study the faults, which can be viewed as time neighboring information. In fact, the inlet temperature T_0 is also an important factor. In summary, we can use exhaust temperatures and their neighboring information along time and space to recognize different faults. If there are n ($n = 13$ in our system) thermometers, we can form a feature vector to describe the state of the exhaust system as

$$F = \left\{ T_0, T_1, T_2, \ldots, T_n, T_1 - T_2, T_2 - T_3, \ldots, \right.$$

$$\left. T_n - T_1, T_1', T_2', \ldots, T_n' \right\},$$

(2)

where $T_{i.}^{d} = T_i(j) - T_i(j-1)$. (j) is the temperature at time j of the ith thermometer.

Apart from the above features, we can also construct other attributes to reflect the conditions of the gas turbine. In this work, we consider a gas turbine with 13 thermometers around the exhaust. So we can form a 40-attribute vector finally.

There are some questions whether all the extracted features are useful for final modeling and how we can evaluate the features and find the most informative features. In fact, there are a number of measures to estimate feature quality, such as dependency in the rough set theory [21], consistency [22], mutual information in the information theory [23], and classification margin in the statistical learning theory [24]. However, all these measures are computed in the original input space, while the effective classification techniques usually implement a nonlinear mapping of the original space to a feature space by a kernel function. In this case, we require a new measure to reflect the classification information of the feature space. Now we extend the traditional information entropy to measure it.

Given a set of samples $U = \{x_1, x_2, ...,\}$, each sample is described with n features $F = \{f_1, f_2, ..., f_n\}$. As to classification learning, each training sample x_i is associated with a decision y_i. As to an arbitrary subset $F \subseteq F$ and a kernel function K, we can calculate a kernel matrix

$$K = \begin{bmatrix} k_{11} & \cdots & k_{1m} \\ \vdots & \ddots & \vdots \\ k_{m1} & \cdots & k_{mm} \end{bmatrix},$$

(3)

where $k_{ij} = k(x_i, x_j)$. The Gaussian function is a representative kernel function:

$$k_{ij} = \exp\left(-\frac{\|x_i - x_j\|^2}{\sigma}\right)$$

(4)

A number of kernel functions have the properties (1) $k_{ij} \in [0, 1]$; (2) $k_{ij} = k_{ji}$.

Kernel matrix plays a bottleneck role in kernel based learning [25]. All the information that a classification algorithm can use is hidden in this matrix. In the same time, we can also calculate a decision kernel matrix as

$$D = \begin{bmatrix} d_{11} & \cdots & d_{1m} \\ \vdots & \ddots & \vdots \\ d_{m1} & \cdots & d_{mm} \end{bmatrix}$$

(5)

Where $d_{ij} = 1$ if $y_i = y_j$; otherwise, $d = 0$. In fact, the matrix D is a matching kernel.

Definition 1. Given a set of samples $U = \{x_1, x_2,...,\}$, each sample is described with n features $F = \{f_1, f_2,...,f_n\}$. $F \subseteq F$, K is a kernel matrix over U in terms of $^{F'}$ .Then the entropy of $^{F'}$ is defined as

$$E(K) = -\frac{1}{m}\sum_{i=1}^{m}\log_2\frac{K_i}{m},$$

(6)

where. $K_i = \sum_{j=1}^{m}=1\, k_{ij}$.

As to the above entropy function, if we use Gaussian function as the kernel, we have $\log_2 m \geq E(K) \geq 0$. $E(K) = 0$ if and only if $k_{ij} = 1$ $\forall i, j$. $E(K) = \log_2 m$ if and only if $k_{ij} = 0$, $i = j$. $E(K) = 0$ means that any pair of samples cannot be distinguished with the current features, while $E(K) = \log_2 m$ means any pair of samples is different from each other. So they can be distinguished. These are two extreme cases. In real-world applications, part of samples can be discerned with the available features, while others are not. In this case, the entropy function takes value in the interval $[0, \log_2 m]$.

Moreover, it is easy to show that if $K_1 \subseteq K_2$, $(K_1) \geq (K_2)$, where $K_1 \subseteq K_2$ means $K_1(xi, x_j) \leq K_2(x_i, x_j)$, $\forall i, j$.

Definition 2. Given a set of samples $U = \{x_1, x_2,...,\}$, each sample is described with n features $F = \{f_1, f_2,...,f_n\}$. $F_1, F_2 \subseteq F$. K_1 and K_2

are two kernel matrices induced by F_1 and F_2. K is a new function computed with $F_1 \cup F_2$. Then the joint entropy of F_1 and F_2 is defined as

$$E(K_1, K_2) = E(K) = -\frac{1}{m}\sum_{i=1}^{m}\log_2\frac{K_i}{m},$$

(7)

where . $K_i = \Sigma_{j=1}^{m}=1 \ k_{ij}$.

As to the Gaussian function, $(x_i, x_j)=K_1(x_i, x_j) \times K_2(x_i, x_j)$. Thus $K \subseteq K_1$ and $K \subseteq K_2$. In this case, $(K) \geq (K_1)$ and $E(K) \geq E(K_2)$.

Definition 3. Given a set of samples $U = \{x_1, x_2,...,\}$, each sample is described with n features $F = \{f_1, f_2,..., f_n\}$.One has F_1, F2 $\subseteq F$. K_1 and K_2 are two kernel matrices induced by F_1 and F_2. K is a new kernel function computed with $F_1 \cup F_2$. Knowning F_1, the condition entropy of F_2 is defined as

$$E(K_1 \mid K_2) = E(K) - E(K_1)$$

(8)

As to the Gaussian kernel, $(K) \geq (K_1)$ and $E(K) \geq E(K_2)$, so $E(K_1 \mid K_2) \geq 0$ and $E(K_2 \mid K_1) \geq 0$.

Definition 4. Given a set of samples $U = \{x_1, x_2,...,\}$, each sample is described with n features $F = \{f_1, f_2,...,f_n\}$. One has F_1, $F_2 \subseteq F$. K_1 and K_2 are two kernel matrices induced by F_1 and F_2. K is a new kernel function computed with $F_1 \cup F_2$. Then the mutual information of K_1 and K_2 is defined as

$$MI(K_1, K_2) = E(K_1) + E(K_2) - E(K)$$

(9)

As to Gaussian kernel, $MI(K_1, K_2) = MI(K_2, K_1)$. If $K_1 \subseteq K_2$, we have $MI(K_1, K_2) = E(K_2)$ and if $K_2 \subseteq K_1$, we have $MI(K_1, K_2) = E(K_1)$.

Please note that if $F1 \subseteq F2$, we have $K2 \subseteq K1$. However, $K2 \subseteq K1$ does not mean $F1 \subseteq F2$.

Definition 5. Given a set of samples $U = \{x_1, x_2,...,\}$, each sample is described with n features $F = \{f_1, f_2,...,f_n\}$. $F' \subseteq F$, K is a kernel matrix over U in terms of , and D is the kernel matrix computed with the decision. Then the feature significance F' related to the decision is defined as

$$\text{MI}(K, D) = E(K) + E(D) - E(K, D)$$

(10)

MI(K, D) measures the importance of feature subset in the kernel space to distinguish different classes. It can be understood as a kernelized version of Shannon information entropy, which is widely used feature evaluation selection. In fact, it is easy to derive the equivalence between this entropy function and Shannon entropy in the condition that the attributes are discrete and the matching kernel is used.

Now we show an example in gas turbine fault diagnosis. We collect 3581 samples from two sets of gas turbine systems. 1440 samples are healthy and the others belong to four kinds of faults: load rejection, sensor fault, fuel switching, and salt spray corrosion. The numbers of samples are 45, 588, 71, and 1437, respectively. Thirteen thermometers are installed in the exhaust. According to the approach described above, we form a 40-dimensional vector to represent the state of the exhaust. Obviously, the classification task is not understandable in such high dimensional space. Moreover, some features may be redundant for classification learning, which may confuse the learning algorithm and reduce modeling performance. So it is a key preprocessing step to select the necessary and sufficient subsets.

Here we compare the fuzzy rough set based feature evaluation algorithm with the proposed kernelized mutual information. Fuzzy dependency has been widely discussed and applied in feature selection and attribute reduction these years [26–28]. Fuzzy dependency can be understood as the average distance from the samples and their nearest neighbor belonging to different classes, while the kernelized mutual information reflects the relevance between features and decision in the kernel space.

Comparing Figures 10 and 11, significant difference is obtained. As to fuzzy rough sets, Feature 5 produces the largest dependency and then Feature 38. However, Feature 39 gets the largest mutual information, and Feature 2 is the second one. Thus different feature evaluation functions will lead to completely different results.

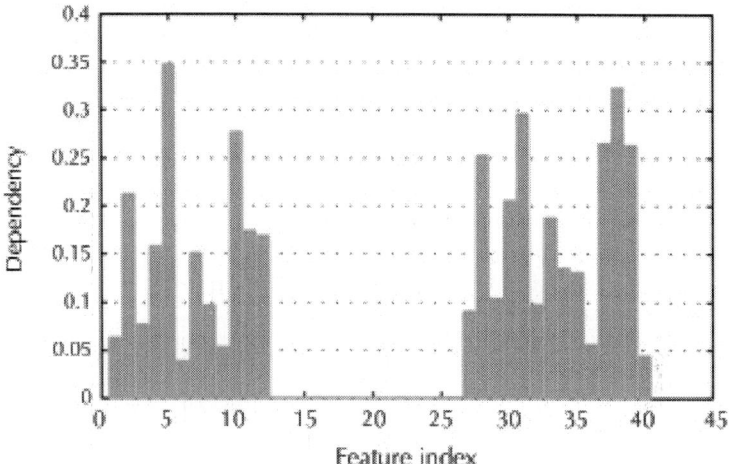

Figure 10: Fuzzy dependency between a single feature and decision.

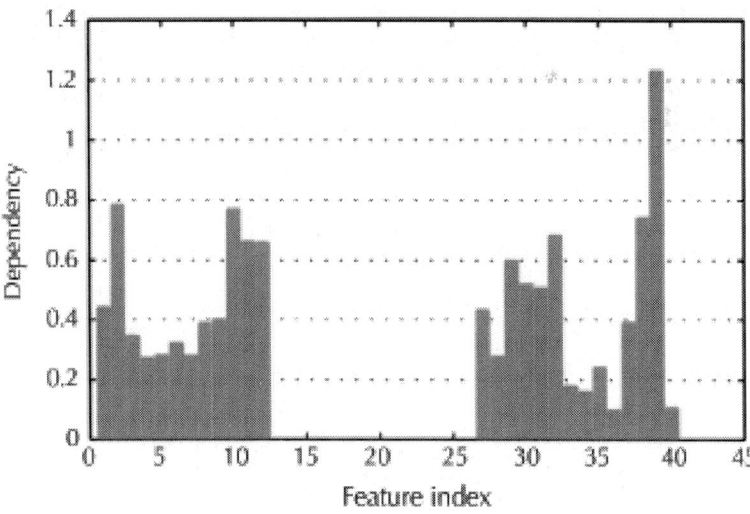

Figure 11: Kernelized mutual information between a single feature and decision.

Figures 10 and 11 present the significance of single features. In applications, we should combine a set of features. Now we consider a greedy search strategy. Starting from an empty set and the best features are added one by one. In each round, we select a feature which produces the largest significance increment with the selected subset. Both fuzzy dependency and kernelized mutual information increase monotonically if new attributes are added. If the selected features are sufficient for classification, these two functions will keep invariant by adding any new attributes. So we can stop the algorithm if the increment of significance is less than a given threshold. The significances of the selected feature subset are shown in Figures 12 and 13, respectively.

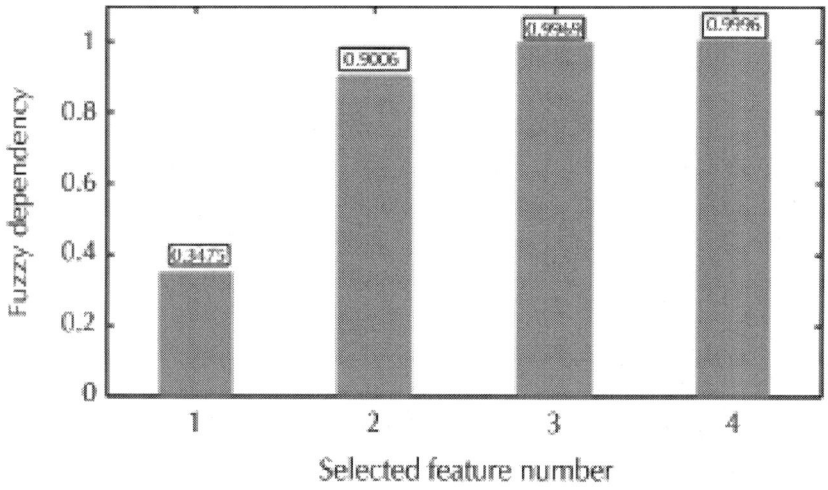

Figure 12: Fuzzy dependency between the selected features and decisions (Features 5, 37, 2, and 3 are selected sequentially).

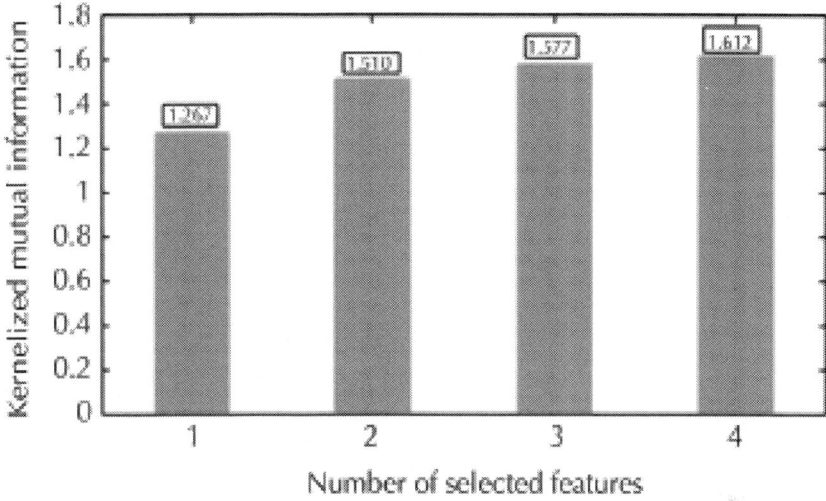

Figure 13: Kernelized mutual information between the selected features and decisions (Features 39, 31, 38, and 40 are selected sequentially).

In order to show the effectiveness of the algorithm, we give the scatter plots in 2D spaces, as shown in Figures 14 to 16, which are expended by the feature pairs selected by fuzzy dependency, kernelized mutual information, and Shannon mutual information. As to fuzzy dependency, we select Features 5, 37, 2, and 3.Then there are 4×4 = 16 combinations of feature pairs. The subplot in the ith row and jth column in Figure 14 gives the scatters of samples in 2D space expanded by the ith selected feature and the jth selected feature.

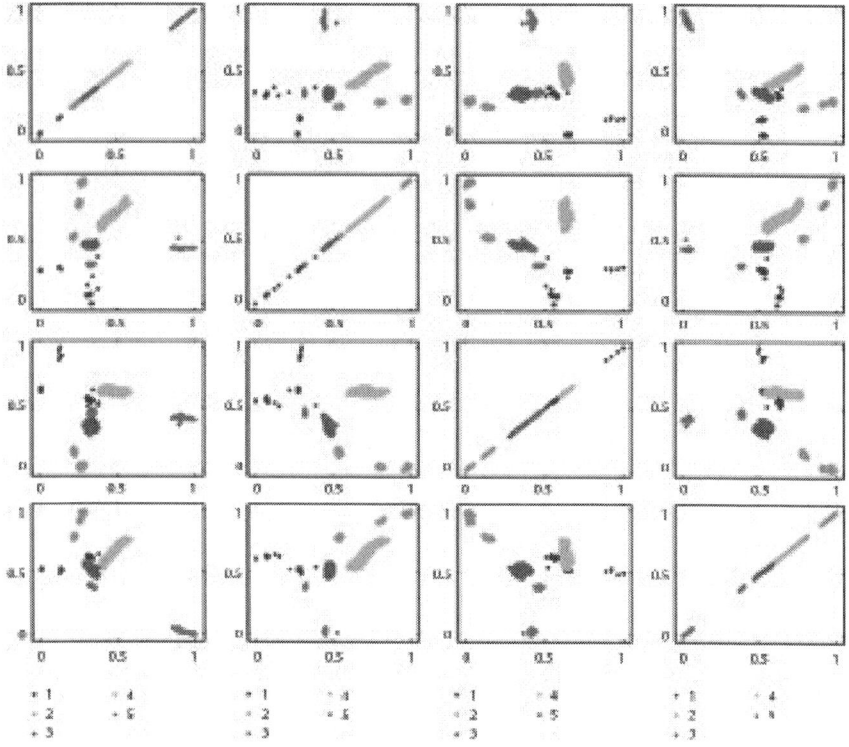

Figure 14: Scatter plots in 2D space expended with feature pairs selected by fuzzy dependency.

Observing the 2nd subplots in the first row of Figure 14, we can find that the classification task is nonlinear. The first class is dispersed and the third class is also located at different regions, which leads to the difficulty in learning classification models.

However, in the corresponding subplot of Figure 15, we can see that each class is relatively compact, which leads to a small intraclass distance. Moreover, the samples in five classes can be classified with some linear models, which also bring benefit for learning a simple classification model.

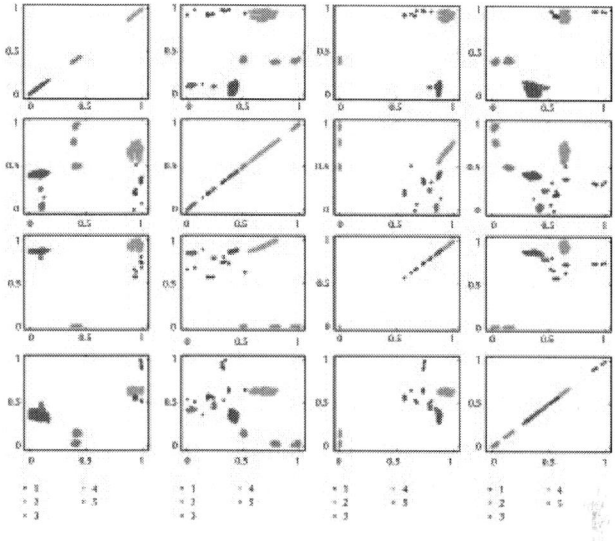

Figure 15: Scatter in 2D space expended with feature pairs selected by kernelized mutual information.

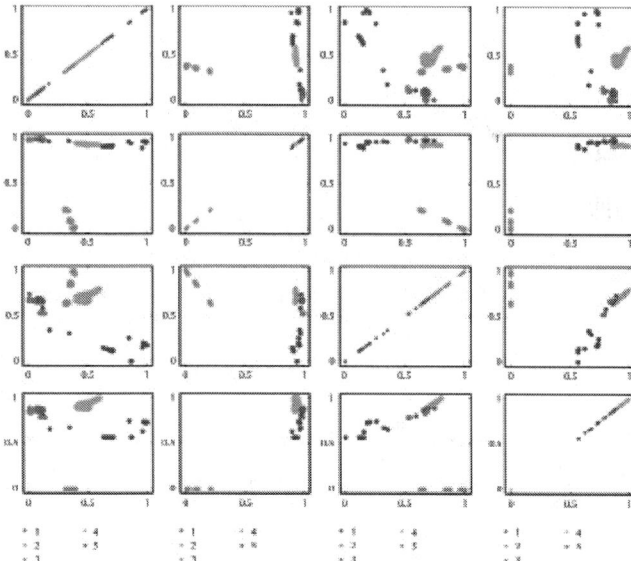

Figure 16: Scatter in 2D space expended with feature pairs selected by Shannon mutual information.

Comparing Figures 15 and 16, we can find that different classes are overlapped in feature spaces selected by Shannon mutual information or get entangled, which leads to the bad classification performance.

DIAGNOSIS MODELING WITH INFORMATION ENTROPY BASED DECISION TREE ALGORITHM

After selecting the informative features, we now go to classification modeling. There are a great number of learning algorithms for building a classification model. Generalization capability and interpretability are the two most important criteria in evaluating an algorithm. As to fault diagnosis, a domain expert usually accepts a model which is consistent with his common knowledge. Thus, he expects the model is understandable; otherwise, he will not believe the outputs of the model. In addition, if the model is understandable, a domain expert can adapt it according to his prior knowledge, which makes the model suitable for different diagnosis objects.

Decision tree algorithms, including CART [29], ID3 [17], and C4.5 [18], are such techniques for training an understandable classification model. The learned model can be transformed into a set of rules. All these algorithms build a decision tree from training samples. They start from a root node and select one of the features to divide the samples with cuts into different branches according to their feature values. This procedure is interactively conducted until the branch is pure or a stopping criterion is satisfied. The key difference lies in the evaluation function in selecting attributes or cuts. In CART, splitting rules GINI and Twoing are adopted, while ID3 uses information gain and C4.5 takes information gain ratio. Moreover, C4.5 can deal with numerical attributes compared with ID3. Competent performance is usually observed with C4.5 in real-world applications compared with some popular algorithms, including SVM and Baysian net. In this work, we introduce C4.5 to train classification models. The pseudocode of C4.5 is formulated as follows.

Decision tree algorithm C4.5

Input: a set of training samples $U = \{x_1, x_2,....,\}$ with features s $F =$

$\{f_1, f_2,...,f_n\}$ Stopping criterion Output: decision tree

- Check for sample set
- For each attribute f compute the normalized information gain ratio from splitting on a
- Let f_best be the attribute with the highest normalized information gain
- Create a decision node that splits on f_best
- Recurse on the sublists obtained by splitting on f_best, and add those nodes as children of node until stopping criterion is satisfied
- Output T.

We input the data sets into C4.5 and build the following two decision trees. Features 5, 37, 2, and 3 are included in the first dataset, and Features 39, 31, 38, and 40 are selected in the second dataset. The two trees are given in Figures 17 and 18, respectively.

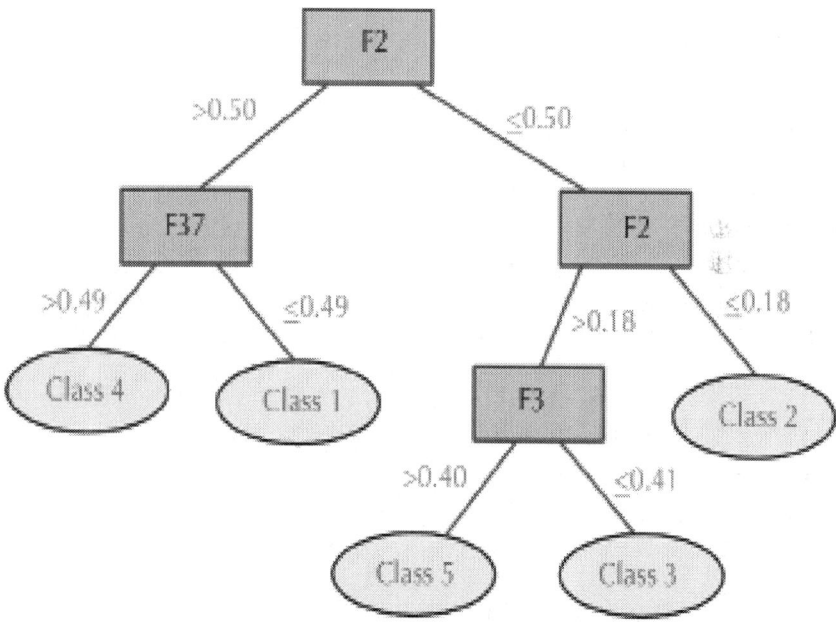

Figure 17: Decision tree trained on the features selected with fuzzy rough sets.

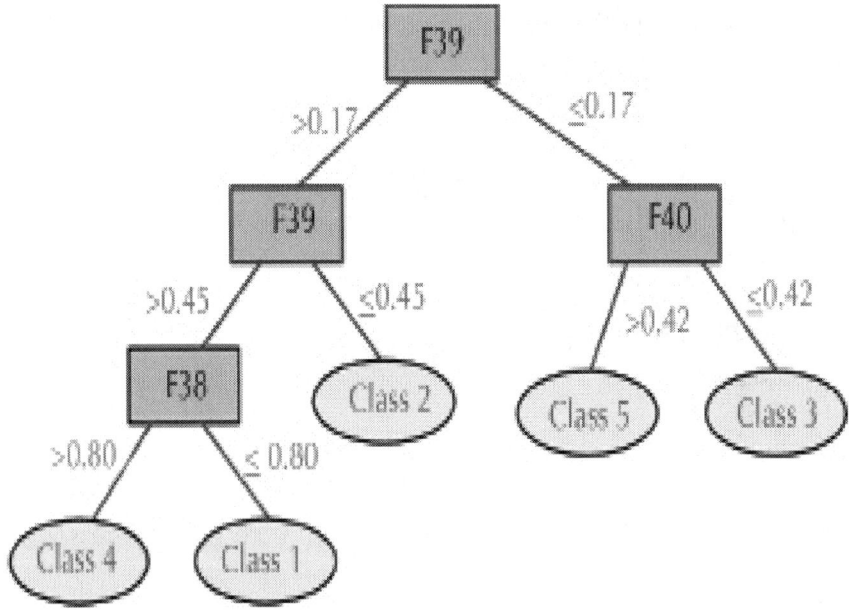

Figure 18: Decision tree trained on the features selected with kernelized mutual information.

We start from the root node to a leaf node along the branch, and then a piece of rule is extracted from the tree. As to the first tree, we can get five decision rules:

- if F2 > 0.50 and F37 > 0.49, then the decision is Class 4;
- if F2 > 0.50 and F37 ≤ 0.49, then the decision is Class 1;
- if 0.18 < F2 ≤ 0.50 and F3 > 0.41, then the decision is Class 5;
- if 0.18 < F2 ≤ 0.50 and F3 ≤ 0.41, then the decision is Class 3;
- if F2 ≤ 0.18, then the decision is Class 2.

As to the second decision tree, we can also obtain some rules as

- if F39 > 0.45 and F38 > 0.80, then the decision is Class 4;
- if F39 > 0.45 and F38 ≤ 0.80, then the decision is Class 1;
- if 0.17 < F39 ≤ 0.45, then the decision is Class 2;
- if F39 ≤ 0.17 and F40 > 0.42, then the decision is Class 5;
- if F39 ≤ 0.17, and F40 ≤ 0.42, then the decision is Class 3.

CONCLUSIONS AND FUTURE WORKS

Automatic fault detection and diagnosis are highly desirable in some industries, such as offshore oil well drilling platforms, for such systems are self-monitoring without man on duty. In this work, we design an intelligent abnormality detection and fault recognition technique for the exhaust system of gas turbines based on information entropy, which is used in measuring the uniformity of exhaust temperatures, evaluating the significance of features in kernel spaces, and selecting splitting nodes for constructing decision trees. The main contributions of the work are two parts. First, we introduce the entropy function to measure the uniformity of exhaust temperatures. The measurement is easy to compute and understand. Numerical experiments also show its effectiveness. Second, we extend Shannon entropy for evaluating the significance of attributes in kernelized feature spaces. We compute the relevance between a kernel matrix induced with a set of attributes and the matrix computed with the decision variable. Some numerical experiments are also presented. Good results are derived.

Although this work gives an effective framework for automatic fault detection and recognition, the proposed technique is not tested on large-scale real tasks. We have developed a remote state monitoring and fault diagnosis system. Large scale data are flooding into the center. In the future, we will improve these techniques and develop a reliable diagnosis system.

ACKNOWLEDGMENTS

This work is partially supported by National Natural Foundation under Grants 61222210 and 61105054.

REFERENCES

1. C. Wang, L. Xu, and W. Peng, "Conceptual design of remote monitoring and fault diagnosis systems,"Information Systems, vol. 32, no. 7, pp. 996–1004, 2007.

2. W. Donat, K. Choi, W. An, S. Singh, and K. Pattipati, "Data visualization, data reduction and classifier fusion for intelligent fault diagnosis in gas turbine engines," Journal of Engineering for Gas Turbines and Power, vol. 130, no. 4, Article ID 041602, 2008.

3. Y. G. Li and P. Nilkitsaranont, "Gas turbine performance prognostic for condition-based maintenance," Applied Energy, vol. 86, no. 10, pp. 2152–2161, 2009.

4. H. Bassily, R. Lund, and J. Wagner, "Fault detection in multivariate signals with applications to gas turbines," IEEE Transactions on Signal Processing, vol. 57, no. 3, pp. 835–842, 2009.

5. K. Young, D. Lee, V. Vitali, and Y. Ming, "A fault diagnosis method for industrial gas turbines using bayesian data analysis," Journal of Engineering for Gas Turbines and Power, vol. 132, Article ID 041602, 2010.

6. B. Yu, D. Liu, and T. Zhang, "Fault diagnosis for micro-gas turbine engine sensors via wavelet entropy," Sensors, vol. 11, no. 10, pp. 9928–9941, 2011.

7. S. Wu, P. Wu, C. Wu, J. Ding, and C. Wang, "Bearing fault diagnosis based on multiscale permutation entropy and support vector machine," Entropy, vol. 14, no. 8, pp. 1343–1356, 2012.

8. S. Wu, C. Wu, T. Wu, and C. Wang, "Multi-scale analysis based ball bearing defect diagnostics using Mahalanobis distance and support vector machine," Entropy, vol. 15, no. 2, pp. 416–433, 2013.

9. H. A. Nozari, M. A. Shoorehdeli, S. Simani, and H. D. Banadaki, "Model-based robust fault detection and isolation of an industrial gas turbine prototype using soft computing techniques,"Neurocomputing, vol. 91, pp. 29–47, 2012.

10. S. Sarkar, X. Jin, and A. Ray, "Data-driven fault detection in aircraft engines with noisy sensor measurements," Journal of Engineering for Gas Turbines and Power, vol. 133, no. 8, Article ID 081602, 10 pages, 2011.

11. C. E. Shannon, "A mathematical theory of communication," The Bell System Technical Journal, vol. 27, pp. 379–423, 1948.

12. S. M. Pincus, "Approximate entropy as a measure of system complexity," Proceedings of the National Academy of Sciences

of the United States of America, vol. 88, no. 6, pp. 2297–2301, 1991.

13. L. I. Kuncheva and C. J. Whitaker, "Measures of diversity in classifier ensembles and their relationship with the ensemble accuracy," Machine Learning, vol. 51, no. 2, pp. 181–207, 2003.

14. I. Csiszár, "Axiomatic characterizations of information measures," Entropy, vol. 10, no. 3, pp. 261–273, 2008.

15. M. Zanin, L. Zunino, O. A. Rosso, and D. Papo, "Permutation entropy and its main biomedical and econophysics applications: a review," Entropy, vol. 14, no. 8, pp. 1553–1577, 2012.

16. C. Wang and H. Shen, "Information theory in scientific visualization," Entropy, vol. 13, pp. 254–273, 2011.

17. J. R. Quinlan, "Induction of decision trees," Machine Learning, vol. 1, no. 1, pp. 81–106, 1986.

18. J. Quinlan, C4.5: Programs for Machine Learning, Morgan Kaufmann, 1993.

19. D. N. Reshef, Y. A. Reshef, H. K. Finucane et al., "Detecting novel associations in large data sets,"Science, vol. 334, no. 6062, pp. 1518–1524, 2011.

20. C. J. C. Burges, "A tutorial on support vector machines for pattern recognition," Data Mining and Knowledge Discovery, vol. 2, no. 2, pp. 121–167, 1998.

21. Q. Hu, D. Yu, W. Pedrycz, and D. Chen, "Kernelized fuzzy rough sets and their applications," IEEE Transactions on Knowledge and Data Engineering, vol. 23, no. 11, pp. 1649–1667, 2011.

22. Q. Hu, H. Zhao, Z. Xie, and D. Yu, "Consistency based attribute reduction," in Advances in Knowledge Discovery and Data Mining, Z.-H. Zhou, H. Li, and Q. Yang, Eds., vol. 4426 of Lecture Notes in Computer Science, pp. 96–107, Springer, Berlin, Germany, 2007.

23. Q. Hu, D. Yu, and Z. Xie, "Information-preserving hybrid data reduction based on fuzzy-rough techniques," Pattern Recognition Letters, vol. 27, no. 5, pp. 414–423, 2006.

24. R. Gilad-Bachrach, A. Navot, and N. Tishby, "Margin based feature selection—theory and algorithms," in Proceedings of the 21th International Conference on Machine Learning (ICML '04), pp. 337–344, July 2004.

25. J. Shawe-Taylor and N. Cristianini, Kernel Methods for Pattern Analysis, Cambridge University Press, 2004.

26. Q. Hu, Z. Xie, and D. Yu, "Hybrid attribute reduction based on a novel fuzzy-rough model and information granulation," Pattern Recognition, vol. 40, no. 12, pp. 3509–3521, 2007.

27. R. Jensen and Q. Shen, "New approaches to fuzzy-rough feature selection," IEEE Transactions on Fuzzy Systems, vol. 17, no. 4, pp. 824–838, 2009.

28. S. Zhao, E. C. C. Tsang, and D. Chen, "The model of fuzzy variable precision rough sets," IEEE Transactions on Fuzzy Systems, vol. 17, no. 2, pp. 451–467, 2009.

29. L. Breiman, J. H. Friedman, R. A. Olshen, and C. J. Stone, Classification and Regression Trees, Wadsworth & Brooks/Cole Advanced Books & Software, Monterey, Calif, USA, 1984.

Surface Temperatures Determination with Influencing Convective and Radiative Thermal Resistance Parameters of Combustor of Gas Turbine

Ebene Ufot[1, 2], Ibiba Emmanuel Douglas[1], and Howel Iberefata Hart[1]

[1]Department of Mechanical Engineering, Rivers State University of Science and Technology, Port Harcourt, Nigeria

[2]Department of Mechanical Engineering, University of Uyo, Uyo, Nigeria

ABSTRACT

Surface temperatures were determined with due consideration of the influencing thermal conditions of conductive, convective and radiative heat. A general condition of heat influx to a point was formulated with the end effect of such influx to the receiving point. It was noted that the heat flow will cause a rate of change of internal energy of the point. Based on the theory of the rate of change of internal energy, a combustor model of cylindrical cross-section was used to generate out the timely temperature equation. Further work was done on this model equation to convert it to nondimensional. The conversion of this equation was very essential in summing up the parameters that can influence the timely generation of the temperatures. Interestingly, it is noted that when a material withstands temperatures, it will equally withstand the thermal stresses that inherently will be developed in it. From the results, the work came up with a table showing the range of these slope figures of equations, a point was also found for a vital recommendation for further studies, where such figures can be used to check the suitability for thermal stress levels and the lifetime of combustor of such thickness.

INTRODUCTION

The temperatures of materials are always critically observed, for a lot of material properties depend on the state of temperatures of the body, viz. the mechanical, thermal, environmental, electrical and chemical. In power plants, like gas turbines, steam turbines or sophisticated nuclear plants, material selection is always a crucial issue, in terms of wall-thickness, temperatures and stress suitability. For example, in Plant Materials.pdf, [1] has it stated that "Aluminium has been ruled out for power reactor application due to hydrogen generation and it does not have adequate mechanical and corrosion-resistant properties at the high operating temperatures". It is further stated that good heat transfer properties are desirable in order that the heat produced will be efficiently transferred. Some authors, [2] -[5] describe the importance of wall-thickness in material selection, but never concluded the issue on determining how suitable the thickness would be. This work is showing how to prove this suitability with facts and figures.

Generally, the way a material reacts to the environment determines its lifetime and even its suitability for further applications. The environments may be mechanical, thermal, chemical, etc. In a thermal environment, the influencing parameters are the convective, the radiative and the conductive. And in [6] [7], the material reaction to the heat inflows is mostly realized in its temperatures, since the internal energy must change. This work is determining the temperatures at the period of the transient thermal loading of the material.

MODELING THE FLOW OF HEAT TO A POINT

Determination of Total Energy to a Point

From Figure 1, considering a point, Node i (m, n) that is thermally not isolated from its neighbouring points, Node j The sum total of heat energy that is transferred into Node i through the individual resistances can be given as:

$$\sum_{j} \frac{Tj^{P} - Ti^{P}}{Rij}$$

(1)

Where m, n, are the space coordinates.

Point i is a singled out point in the space.

Point j is any point in the neighborhood that is connected to point i thermally using conducting, but not heatgenerating rods.

Rij is the thermal resistances of such fictitious rods allowing the heat to flow from point j to point i. Tj^{P} denotes the temperature at point j at the point in time P; Tj^{P} is assumed to be greater than Ti^{P}; (for coding purposes, hereafter, typed T_{jp})

Ti^{P} denotes the temperature at point I, (coded T_{ip}) Ti^{P1}: denotes the temperature at point i, a short interval of time after the point must have received the heat energy from its surrounding points (i.e. heat transfer either by radiation, convection or conduction), coded $T_{i(P+1)}$; ρi is the density of the node point i; ci is the specific heat capacity.

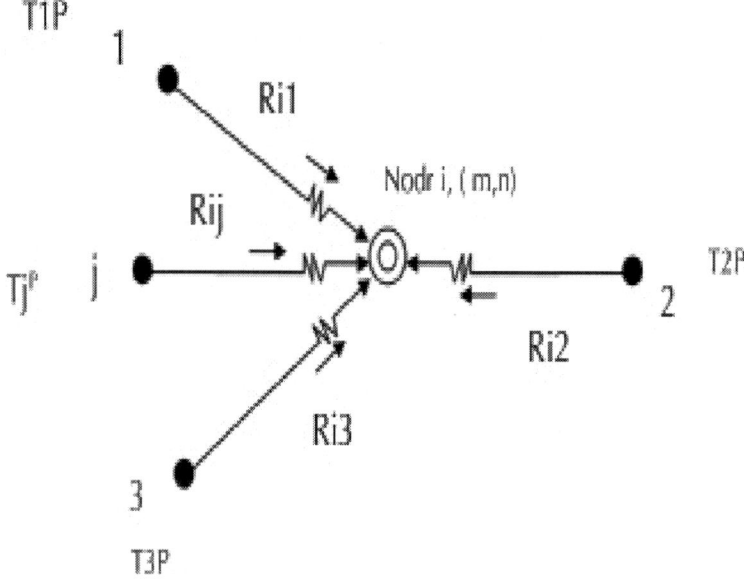

Figure 1: Inflow of heat to a node point.

DVi is the elemental volume. Ci is the heat capacity where Ci = ρi ci DVi; Dt is the time interval.

DE is the increase in the internal energy of the element; qi is the generated heat, if any, at point i.

Setting the sum of the energy (conducted, convicted or radiated) into the node equal to the increase in the internal energy of the node, this energy-intake by the point is indicated by the temperature rise within the short interval of time.

In a steady-state, the net sum is zero, i.e. the net energy transfer is zero, while for the unsteady-state problems of our interest, here, the net energy-transfer into the node must be evidenced as an increase in internal energy of the element.

$$\text{Thus:} \quad \frac{\Delta E}{\Delta \tau} = \rho c \Delta V_i \frac{Ti^{P1} - Ti^{P}}{\Delta \tau} \tag{2}$$

So that the general resistance-capacity formulation [8], on the node, is expressed as:

$$qi + \sum_j \frac{Tj^P - Ti^P}{Rij} = Ci \frac{Ti^{P1} - Ti^P}{\Delta \tau}$$

(3)

Under steady state condition Equation (3) becomes:

$$qi + \sum_j \frac{Tj^P - Ti^P}{Rij} = 0$$

(4)

MODELING THE TEMPERATURE GENERATION AT A NODE POINT

In a particular combustor model (Figure 2), of the following dimensions: External diameter = 72 cm, Wall

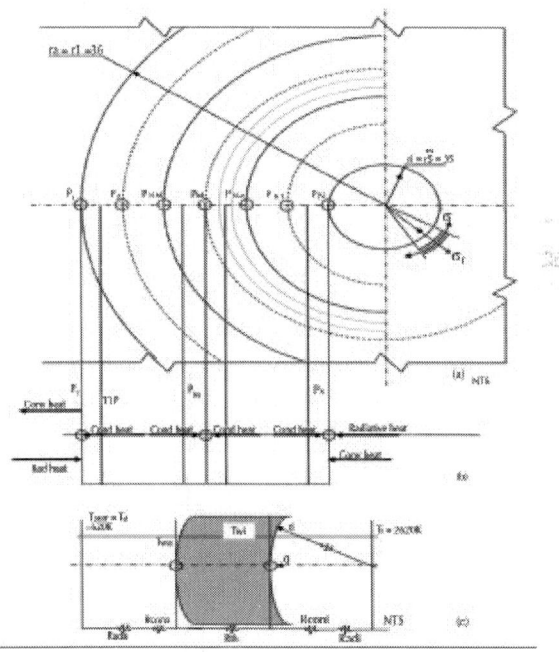

Figure 2: The cross-section of combustor. ri = inside radius of cylinder. ra = center of cylinder to outside surface.

Thickness = 1 cm, and total length of 100 cm, this work is concerned with node point 1, P_1, how its temperature is changing during the period of the transient thermal loading of the model combustor.

The compressor discharges directly to the combustor, where initially the bulk discharge volume is designed to divide into two streams: the greater stream goes into the annular space between the liner walls and the casing. This annular space can be equally designed as to determine the flow velocity, and thereby the convective heat transfer. With reference to Figure 2(c) the bulk stream temperature enveloping the liner, say temperature surrounding, T_{surr} = 620 K.

The radiative heat resistances for inside and outside bulk streams are noted as Radi and Rada respectively. The heat capacity, C_1:

$$C_1 = \rho * 2\pi r_{m1} * \frac{\Delta r}{2} * c = \rho \pi r_m \Delta r * c$$

(5)

r_m = radial position to center of nodal element; r_m = conductive thermal resistance in material.

$\Delta\varphi$ = polar angular measurement: $\Delta\varphi$ = 2π, by full circular measurement.

Δz = incremental axial distance: Δz = 1 m, by full axial measurement.

ΔV = elemental volume, expressed in [μm^3]. Δr = the redial increment, redial thickness of the element, = density of the material

Parameter Information for a Nodal Equation

We assume on the outer surface a variable heat convection coefficient, h_a.

$$h_a = h * \Delta T^{1/4} \left[\mathrm{W/m^2 \cdot {}^\circ C} \right]$$

(6)

Where h = a chosen constant coefficient: (h-1.92)

We also assume in the model material,

$\Delta T = T_{surr} - T_{1P}$ i e the temperature difference between that of the surrounding and node point 1

h_a is chosen to vary as the temperature difference across the node at the transient time interval.

$k = 22 \dfrac{W}{m}$ °C, the conduction heat coefficient.

So that the conductive resistance, R_{mi}:

At boundary points,

$$Rm_{N-} = \frac{\Delta r/2}{\left[r_{m_N} - \Delta r/4 \right] \Delta \varphi * \Delta z * k} \quad \text{And,} \quad Rm_{N+} = \frac{\Delta r/2}{\left[r_{mN} + \Delta r/4 \right] \Delta \varphi \Delta z * k}$$

(7)

Rm_{n-} is the conductive thermal resistance for the external nodal point.

Temperature Equation on Outside Wall-Node Point 1

Temperature equation for Boundary point, Node 1:

Generally, the change in internal energy of a nodal element is given by

$$m \cdot c \cdot \Delta T = \rho \Delta V c \cdot \Delta T$$

And the rate of change of internal energy, Where $= Vc \cdot T \Delta \tau$, Where $\Delta \tau$ the transient time interval in seconds. The instantaneous inflow of heat energy to a nodal element equals to the rate of change of internal energy.

For boundary point, Node 1: Analyzing the thermal resistances and balancing the transient energies:

Resulting

$$\sigma \varepsilon A_1 \left[T_{surr}^2 + T_{1P}^2 \right] * \left[T_{surr} + T_{1P} \right] * \left[T_{surr} - T_{1P} \right]$$
$$+ h_a A_1 \Delta T^{1/4} * \left[T_{1P} - T_{surr} \right] + \frac{1}{R_{m1-}} \left[T_{2P} - T_{1P} \right] = \frac{C_1}{\Delta \tau} \left[T_{1(P+1)} - T_{1P} \right]$$

(8)

Where T_{surr} = the bulk temperature of the surrounding, i.e. the peripheral Air stream temperature.

T_{1P} = the nodal temperature before inflow of energies, $T_{1(P+1)}$ = the instantaneous new arrived temperature of the node, after receiving the inflow of energies.

Expanding Equation (8),

$$
\therefore T_{1(P+1)} = \frac{\Delta\tau}{C_1}\left[\sigma\varepsilon A_1\left(T_{urrs}^2 + T_{1P}^2\right)\left(T_{surr} + T_{1P}\right)T_{surr} + h_a A_1\left(T_{1P} - T_{surr}\right)^{1/4}T_{surr} + \frac{1}{R_{m1-}}T_{2P}\right]
$$
$$
+ \left\{1 - \frac{\Delta\tau}{C_1}\left[\sigma\varepsilon A_1\left(T_{surr}^2 + T_{1P}^2\right)\left(T_{surr} + T_{1P}\right) + h_a A_1\left(T_{1P} - T_{surr}\right)^{1/4} + \frac{1}{R_{m1-}}\right]\right\}T_{1P}
$$

(9)

Equation (9) is the surface temperature equation.

Parametric Temperature Curves-Buckingham (Theorem)

Equation (9), the surface temperature equation can be further expressed in a non-dimensional form by dividing the whole function by the surrounding temperature, T_{surr}:

That is,

$$
\frac{T_{1(P+1)}}{T_{surr}} = \frac{\Delta\tau}{C_1}\left[\sigma\varepsilon A_1 * \left(T_{surr}^2 + T_{1P}^2\right)\left(T_{surr} + T_{1P}\right) * \frac{T_{surr}}{T_{surr}}\right.
$$
$$
\left. + h_a A_1\left(T_{surr} - T_{1P}\right)^{0.25} * \frac{T_{surr}}{T_{surr}} + 1/R_{m1-} * \frac{T_{2P}}{T_{surr}}\right]
$$
$$
+ \left\{1 - \frac{\Delta\tau}{C_1}\left[\sigma\varepsilon A_1 * \left(T_{surr}^2 + T_{1P}^2\right)\left(T_{surr} + T_{1P}\right)\right.\right.
$$
$$
\left.\left. + h_a A_1\left(T_{surr} - T_{1P}\right)^{0.25} + 1/Rm_{1-}\right]\right\} * \frac{T_{1P}}{T_{surr}}
$$

(10)

Expressing all Independent Parameters in Equation (10), the Buckingham π Theorem can be applied to give result in a non-dimensional function, as:

$$\frac{T_{1(P+1)}}{T_{surr}} = \left[B_{Rad} + B_{Conv} + C_a\right] + \left[1 - \left(B_{Rad} + B_{Conv} + B_{cond}\right)\right] * B \tag{11}$$

Where

$$C_a = B_{cond} * \frac{T_{2P}}{T_{surr}} \tag{12}$$

And

$$B = \frac{T_{1P}}{T_{surr}} \tag{13}$$

With

$$B_{cond} = \frac{\Delta\tau}{C_1} * \frac{1}{R_{m1-}} \tag{14}$$

$$B_{conv} = \frac{\Delta\tau}{C_1} * hA_1, \quad \text{where } h = h_a \Delta T^{0.25} \tag{15}$$

And

$$B_{Rad} = \frac{\Delta\tau}{C_1} * \frac{1}{Rada} \tag{16}$$

Calculating the average heat transfer parameters

$$B_{cond} = Tau * 1/Rm_{1-}; \quad B_{conv} = Tau * h_a * A_1 + \left(T_{av} - T_{surr}\right)^{1/4}$$
$$B_{rad} = Tau * \sigma * \varepsilon * A_1 * \left(T_{av}^2 + T_{surr}^2\right) + \left(T_{av} + T_{surr}\right) \tag{17}$$

Numerically and typically, for Var 1.1, i.e. model combustor wall thickness of 2.5 mm B_{cond} = 0.8714 = constant, B_{conv} = 0.0001 = constant, B_{rad} = 0.0037 = constant,

$$T_{1(P+1)}/T_{surr} = \left[0.0038 + C_a\right] + 0.1248 * B$$

$$(18)$$

Graphical presentation:

Temperature values (non-dimensional) $T_{1(P+1)}/T_{surr}$ can be seen in the results of Program EU405TIPI for all Variants.

RESULTS AND DISCUSSIONS

Non-Dimensional Outer Wall Surface Temperatures

The external node point temperature equation (wall surface) was expressed in a non-dimensional parametric form (by dividing through by the surrounding temperature, T_{surr}, and applying Buckingham p theorem). Listing out the independent parameters involved, a temperature functional relation was expressed as:

$$\frac{T_{1(P+1)}}{T_{surr}} = \left[B_{rad} + B_{conv} + C_a\right] + \left\{1 - \left[B_{rad} + B_{conv} + B_{cond}\right]\right\} * B$$

$$(19)$$

Using Equation (10): This is the required surface temperature equation (non-dimensional).

In Figure 3, the results for computing non-dimensional outer wall surface temperatures for Var 4.1 (2.5 mm wall thickness) are shown. This non-dimensional expression involved the different modes of heat transfer, affecting the outer nodal point. Since it is an intrinsic line, Figure 4 shows the outer wall temperature at any particular time of C-variation.

Figure 5 shows the non-dimensional outer wall surface temperatures as plotted against the non-dimensional temperature parameter, C_a (Var 4.1). A determining line was found in order to make use of the graphical presentation of outer wall surface temperatures. This line was determined and worked out in a computer program, (EU405T1P1).

Other results are given in Appendix. As an intrinsic line, this line turns out to be very important as it fixes the temperature at the instance of C_a variation. In Table 1, equation of change of outer wall surface temperatures was shown against the individual wall thickness.

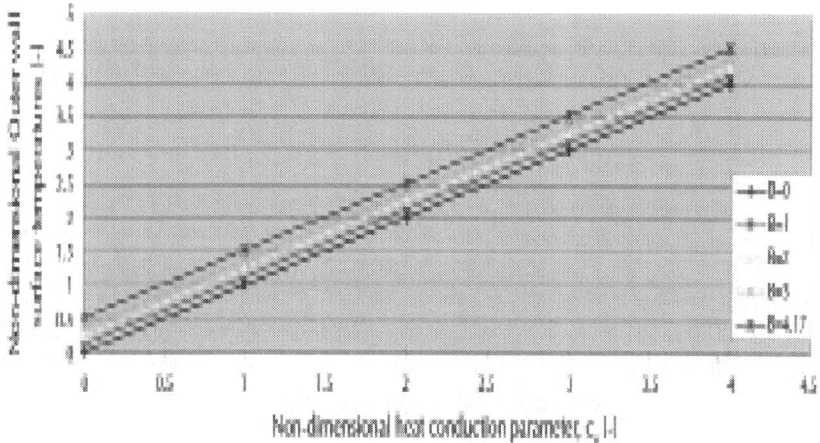

Figure 3: Outer wall surface temperatures (non-dimensional) versus non-dimensional heat conduction parameter C_a (Var 4.1), B = temperature ratio parameter: T_{1p}/T_{surr}.

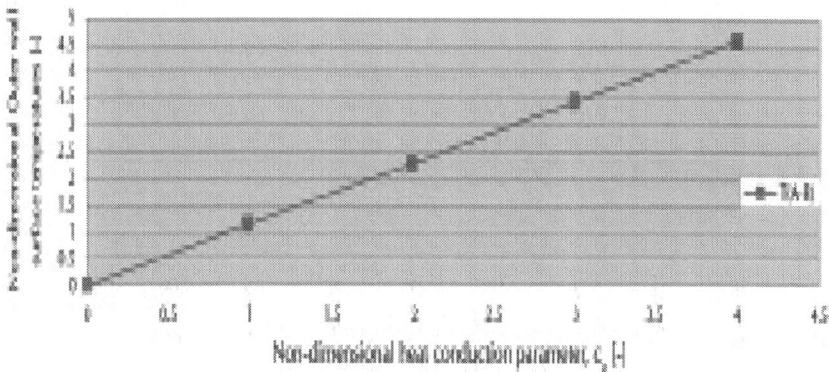

Figure 4: Variation of non-dimensional outer wall temperatures with heat conduction parameter, C_a, for wall thickness of 2.5 mm (Var 4.1).

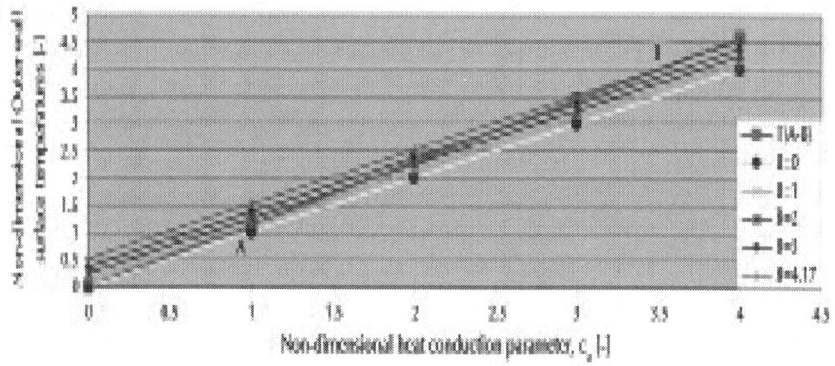

Figure 5: Outer wall surface temperatures (non-dimensional) versus non-dimensional temperature parameter C_a (Var 4.1)- showing lines of B = constant as straight lines. The intrinsic determinant line, AB is shown cutting across the parallel line graphs.

Table 1: Variation of equations of the change-rate of outer wall surface temperatures with non-dimensional temperature parameter, C_a (by Wall thickness)

Var. Nr	Wall thickness [mm]	Equation of Change of outer surface temperaturesTA-B[−]
1	10	$y = 1.265C_a + 0.0171$
2	5	$y = 1.138C_a + 0.0087$
3	3	$y = 1.076C_a + 0.0052$
4.1	2.5	$y = 1.143C_a + 0.0044$
4.2	2.5	$y = 1.129C_a + 0.0166$
5	2	$y = 1.067C_a + 0.005$

CONCLUSIONS

The work has shown the method of generating wall-surface temperatures. Equation (10) generates in ordinary values, while Equation (18) generates non-dimensional values. In achieving this development, it has involved all the infuencing thermal conditions of the conductive, convective and the radiative heat transfers. The work has further shown that temperature generation follows an intrinsic trend line that is distinct to the wall-thickness concerned. Knowing the range of temperatures in application, this method can be recommended for use in confirming the suitability of a material when it will be put in application.

REFERENCES

1. Gilbert Gedeon, P.E. (2013) Nuclear Plant Material Selection and Application. Course No: T05-001 Credit: 5 PDH Continuing Education and Development, Inc. 9 Greyridge Farm Court Stony Point, New York.

2. Yun, N., Jeon, Y.H., Kim, K.M., Lee, D.H. and Cho, H.H. (2009) Thermal and Creep Analysis in a Gas Turbine Combustion Liner. Proceedings of the 4th IASME/WSEAS International Conference on Energy & Environment, Cambridge, November 2009, 315-320.

3. Jayakody, S. (2009) Why Is Selection of Engineering Materials Important? http://www.brighthubengineering.com

4. Ufot, E. (2013) Modeling Thermal Stresses in Combustor of Gas Turbines. Ph.D. Thesis, Rivers State University of Science and Technology, Port Harcourt.

5. Tinga, T., van Kampen, J.F., de Jager, B. and Kok, J.B.W. (2007) Gas Turbine Combustio Liner Life Assessment Using a Combined Fluid/Structural Approach. Thermal Engineering, Twente University, Enschede.

6. Elsner, N. (1975) Grundlagen der Technischen Thermodynamik. 3rd Edition, Akademie-Verlag, Berlin.

7. Hart, H.I. (2005) Engineering Thermodynamics, a First Course. King Jovic Int'l, Port Harcourt.

8.   Holman, J.P. (1997) Heat Transfer. 8th Edition, McGraw-Hill, Inc., New York.

Chapter 7

Dynamic Time-delay Characteristics and Structural Optimization Design of Marine Gas Turbine Intercooler

Ning-bo Zhao[1], Xue-you Wen[1,2], and Shu-ying Li[1]

[1]College of Power and Energy Engineering, Harbin Engineering University, Harbin 150001, China

[2]Harbin Marine Boiler and Turbine Research Institute, Harbin 150078, China

ABSTRACT

Aiming at the rapid mobility of marine gas turbine and the dynamic time-delay problem of intercooler for intercooled cycle marine gas turbine, the dynamic simulation model of intercooler was set up based on effectiveness-number of transfer units (ε-NTU) and lumped parameter method in this paper. The model comprehensively considers related physical properties dependent on temperature. Dynamic response characteristics of gas outlet temperature and pressure and coolant outlet temperature of intercooler with different materials and coolants in the change of operation condition of marine gas turbine were analyzed in detail. Besides, this paper explored the use of simulated annealing algorithm for structural optimization of intercooler. The results showed that both material and coolant were the significant factors that affected the heat transfer and dynamic performance of intercooler. The heat transfer and dynamic performance of the intercooler obtained by using simulated annealing algorithm were better than those of preliminary design.

INTRODUCTION

In recent years, high-power (more than 25 MW) marine gas turbine has aroused the attention of every country [1]. Intercooled (IC) cycle or intercooled regenerated (ICR) cycle technology is a feasible method to develop high-power marine gas turbine, which may represent a major development tendency for a new generation of marine main propulsion plants [2]. The successful application of WR-21 marine gas turbine further verified the feasibility and effectiveness of intercooled regenerated cycle technology to increase engine power for marine gas turbine [3].

As an important part of intercooled or intercooled regenerated cycle marine gas turbine, intercooler will directly influence the performance of engine. The existence of intercooler can aggravate the time-delay characteristics of fluid flow and heat transfer, which make it difficult to match the thermodynamic parameters and develop appropriate control strategies for gas turbine system, and then may influence the maneuverability performance of ship.

In the past half century, many scholars have carried out extensive research from different aspects and made some achievements of both theory and practice about intercooled or intercooled regenerated cycle gas turbine. The research activities mainly focused on intercooled or intercooled regenerated cycle gas turbine performance analysis [4–10], numerical simulation on flow and heat transfer of intercooler

[11–13], and thermodynamic design and optimization of intercooled system [14–16]. Li et al. [4–6] studied the dynamic behaviors and flow parameters optimization of intercooled cycle marine gas turbine based on simulation method. Through the simulation study of fuel supply rate, they obtained the best fuel supply rate curve of intercooled cycle marine gas turbine. Based on the finite time thermodynamic theory, the power and efficiency of an open or closed cycle intercooled gas turbine power plant were analyzed and optimized by adjusting the low pressure compressor inlet relative pressure drop, the mass flow rate, and the distribution of pressure losses along the flow path by Wang et al. [7–10]. According to the actual operation condition of intercooled system for a certain intercooled cycle marine gas turbine, Dong et al. [11–13] made a preliminary thermodynamic analysis and design by using effectiveness-number of transfer units (ε-NTU) method and analyzed the influence of intercooler structural parameters and coolant parameters on the steady-state performance of intercooler. On the base of structure features of plain fin rectangular channels, they also investigated the coupled characteristics of fluid flow and heat transfer by using numerical simulation approach. Through numerical computing, Li et al. [14] and Dong et al. [12] obtained the pressure distribution, temperature distribution, and velocity distribution of intercooler entire flow path, which provided beneficial reference to design and apply intercooler. Xiao [15] discussed the effects of intercooler structure parameters on its flow and heat transfer performance based on effectiveness-number of transfer units (ε-NTU) method. Then they analyzed the dynamic performance and thermal inertia of intercooler under varying operation conditions. Their research results showed that intercooler had the obvious thermal inertia characteristics, which could directly affect the accuracy and reliability of the control system. Besides, Zhang [16] established the dynamic simulation model by using the lumped parameter method and analyzed the dynamic response of gas outlet temperature of intercooler under varying operation conditions. Their conclusions were consistent with the results that have been reported in literature [15]. They also found that heat exchanger weight had a significant influence on the dynamic performance of intercooler.

From the literatures described above, very few studies have been performed on the thorough discussion of relevant factors that could affect the dynamic time-delay characteristics of intercooler. There were also very few studies which discussed the structural optimization design of intercooler using artificial intelligent algorithm. Besides, the dynamic simulation analyses in the existing literatures only considered the change of gas inlet temperature of intercooler, which belonged to the single variable perturbation analysis. However, all the gas inlet temperature, pressure, and mass flow rate of intercooler will change in practical applications. Therefore, it is meaningful to study the dynamic behaviors of intercooler with multivariable perturbation.

The objectives of this study are to investigate the flow and heat transfer coupling dynamic time-delay problem and structural optimization design of intercooler for intercooled cycle marine gas turbine. In this study, the dynamic simulation model of intercooler is modeled based on effectiveness-number of transfer units (ε-NTU) and lumped parameter method. In order to improve the modeling accuracy, temperature dependent thermophysical properties are taken into account due to large temperature differences in the intercooler. Then, three different types of substrate materials (copper, alumina, and copper-nickel alloy) and two different coolants (water and ethylene glycol) are considered to investigate the effect of substrate materials and coolants on dynamic performance of intercooler in detail. Finally, a simulated annealing (SA) algorithm based optimization technique for intercooler will be developed, which minimizes the total weight of intercooler under given space and performance restrictions.

INTERCOOLER PROBLEM DESCRIPTION AND THERMODYNAMIC MODEL

Problem Description of Intercooled System

As the key component, intercooled system is located between the low pressure compressor and high pressure compressor. The intercooler system lowers the high pressure compressor inlet temperature, which can reduce the power consumption of high pressure compressor and increase the power output of the whole marine gas turbine system.

The principle diagram of intercooled cycle marine gas turbine is shown in Figure 1. Intercooler system is composed of two parts, namely, on-engine intercooler (plate-fin heat exchanger) and off-engine intercooler (plate heat exchanger). The on-engine intercooler is designed to decrease the temperature of the pressurized gas coming out of the low pressure compressor. Then the coolant passes through off-engine intercooler, which transfers heat from the coolant to the seawater in order to reduce the temperature of coolant. Compared with off-engine intercooler, on-engine intercooler is more important for marine gas turbine because it has a direct influence on the gas turbine performance.

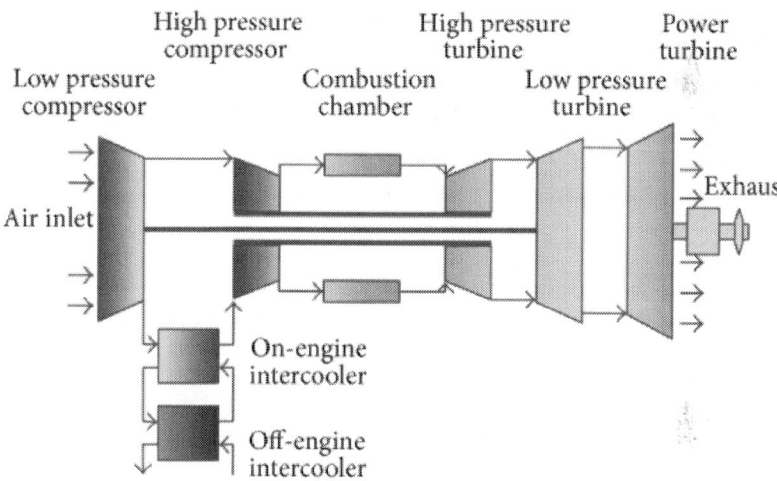

Figure 1: Principle diagram of intercooled cycle marine gas turbine.

According to the requirements of intercooled cycle, three performance parameters should be considered for on-engine intercooler. (1) Reasonable pressure drop of the fluid (especially for gas), including the pressure drop of inlet channels, core part, and outlet channels. When the pressure drop of gas is too high, it will give an additional burden on the high pressure compressor to keep a constant pressure ratio, which means more fuel consumption. (2) Reasonable thermal efficiency of intercooler, which can effectively lower outlet temperature of gas under given space. (3) Reasonable weight of intercooler. There may have a greater thermal inertia when the weight is heavier. Therefore, how to realize the optimization design of high efficiency compact intercooler is one of the important problems for intercooled system.

Due to the complexity in the structure of the intercooler, the detailed information of the fluid flow and heat transfer in intercooler can scarcely be described. Different substrate materials and coolants have different heat transfer performance, which can influence the flow and heat transfer performance of intercooler. So the effects of different materials and coolants on dynamic performance of intercooler are worth studying in detail.

Thermal Modeling of Intercooler

Figures 2 and 3 depict a schematic view of a reverse flow plate-fin intercooler with straight fins and the basic geometric structure of fins, respectively. The following assumptions will be used for the analysis.

- The number of fin layers for the gas side (N_a) is assumed to be one more than the coolant side (N_b). That is, $N_a = N_b + 1$.
- Intercooler works in a steady-state condition.
- The thickness of all the fins is assumed uniform and the thermal resistance can be negligible because the thickness is too small.
- All the parts are made of same material.
- The influences of fouling and corrosion are neglected.

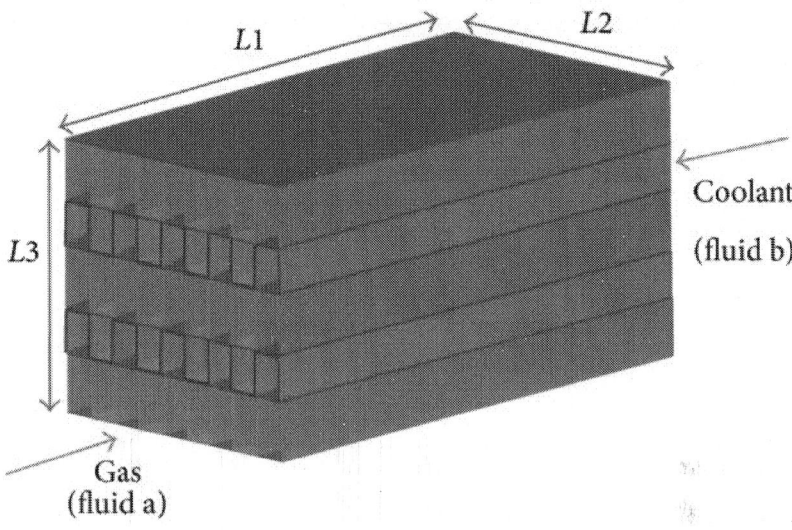

Figure 2: Schematic representation of reverse flow plate-fin intercooler.

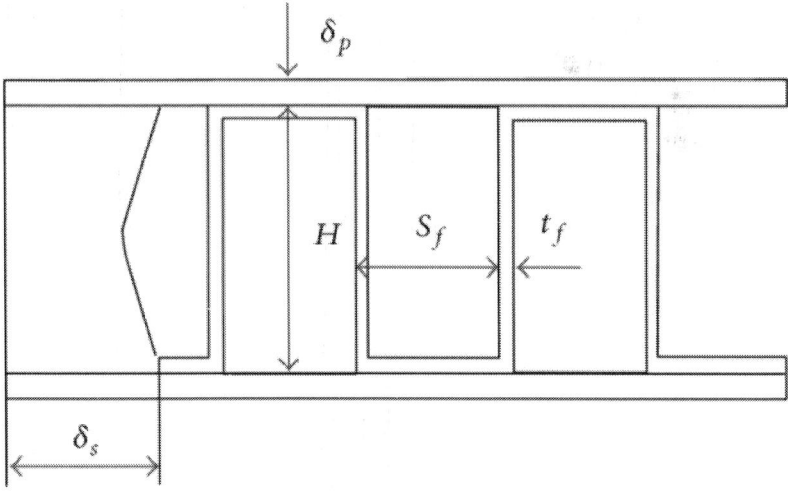

Figure 3: Detailed view of straight fin.

In this study, since the output temperature of the fluids is unspecified, the ε-NTU method is used to assess the flow and heat transfer performance of intercooler in the modeling process. The effectiveness of reverse flow plate-fin intercooler is proposed as

$$\varepsilon = \frac{1 - \exp\left[-NTU\left(1 - Cr\right)\right]}{1 - Cr\exp\left[-NTU\left(1 - Cr\right)\right]},$$

(1)

where $Cr = C_{min} / C_{max}$ is heat capacity ratio. NTU is the number of transfer units. Considering the thermal resistance of the walls, NTU can be determined by

$$\frac{1}{NTU} = \frac{C_{min}}{UA},$$

$$\frac{1}{UA} = \frac{1}{\eta_{ef,a}h_aA_a} + \frac{\delta_p}{\lambda_pA_p} + \frac{1}{\eta_{ef,b}h_bA_b}.$$

(2)

Generally, the flow state of fluids can affect the convective heat transfer coefficient. For the fluids in the fully developed turbulent flow and transition region, the heat transfer coefficient is calculated in terms of Gnielinski [18] which is given as

$$h = \frac{Nu\lambda}{D}.$$

(3)

Nusselt number and hydraulic diameter can be calculated as follows:

$$Nu = \frac{(f/2)\,(Re - 1000)\,Pr}{1 + 12.7(f/2)^{0.5}\,(Pr^{2/3} - 1)},$$

$$D = \frac{2\left(H - t_f\right)\left(S_f - t_f\right)}{H + S_f - 2t_f}.$$

(4)

Reynolds number and Prandtl number are defined as below:

$$Re = \frac{\rho u D}{\mu},$$

$$Pr = \frac{C_p \mu}{\lambda}.$$

(5)

And the Fanning factor f is given as

$$f = \frac{1}{(1.58 \ln Re - 3.28)^2}.$$

(6)

For the fluids in the fully developed laminar flow, the heat transfer coefficient is calculated in terms of Colburn factor [19, 20] which is given as

$$h = \frac{j G C_p}{Pr^{2/3}}.$$

(7

Mass flow velocity can be obtained as follows:

$$G = \frac{W}{A_{ff}}.$$

(8)

The Fanning factor f and Colburn factor j are given as

$$f = \exp\left[0.106566(\ln \text{Re})^2 - 2.12158\,(\ln \text{Re}) + 5.82505\right],$$

$$j = \exp\left[0.103109(\ln \text{Re})^2 - 1.91091\,(\ln \text{Re}) + 3.211\right].$$

$$\text{(9)}$$

In this study, the effective circulation area for the two sides is formulated as

$$A_{ff,a} = \frac{N_a\,(L2 - 2\delta_s)\left(H_a - t_{f,a}\right)\left(S_{f,a} - t_{f,a}\right)}{S_{f,a}}$$

$$A_{ff,b} = \frac{N_b\,(L2 - 2\delta_s)\left(H_b - t_{f,b}\right)\left(S_{f,b} - t_{f,b}\right)}{S_{f,b}}.$$

$$\text{(10)}$$

The heat transfer areas of intercooler for the two sides are calculated by

$$A_a = 2N_a L1\,(L2 - 2\delta_s)\left[1 + \frac{\left(H_a - 2t_{f,a}\right)}{S_{f,a}}\right],$$

$$A_b = 2N_b L1\,(L2 - 2\delta_s)\left[1 + \frac{\left(H_b - 2t_{f,b}\right)}{S_{f,b}}\right].$$

$$\text{(11)}$$

Heat transfer efficiencies of heat transfer surface for the two sides are obtained by

$$\eta_{ef,a} = \frac{\left(S_{f,a} - t_{f,a}\right) + \eta_{f,a}\left(H_a - t_{f,a}\right)}{S_{f,a} + H_a - 2t_{f,a}},$$

$$\eta_{ef,b} = \frac{\left(S_{f,b} - t_{f,b}\right) + \eta_{f,b}\left(H_b - t_{f,b}\right)}{S_{f,b} + H_b - 2t_{f,b}}.$$

$$\text{(12)}$$

Heat transfer efficiencies of fin for the two sides are formulated as

$$\eta_{f,a} = \frac{\tan\left(0.5 m_a H_a\right)}{0.5 m_a H_a},$$

$$\eta_{f,b} = \frac{\tan\left(0.5 m_b H_b\right)}{0.5 m_b H_b}.$$

$$(13)$$

Fin factors for the two sides are defined as follows:

$$m_a = \sqrt{\frac{2 h_a}{\lambda_{f,a} t_{f,a}}},$$

$$m_b = \sqrt{\frac{2 h_b}{\lambda_{f,b} t_{f,b}}}.$$

$$(14)$$

Therefore, the effective heat transfer area for the two sides can be calculated by

$$A_{ef,a} = 2 N_a L1 \left(L2 - 2\delta_s\right) \left[\frac{\left(S_{f,a} - t_{f,a}\right) + \eta_{f,a}\left(H_a - t_{f,a}\right)}{S_{f,a}}\right],$$

$$A_{ef,b} = 2 N_b L1 \left(L2 - 2\delta_s\right) \left[\frac{\left(S_{f,b} - t_{f,b}\right) + \eta_{f,b}\left(H_b - t_{f,b}\right)}{S_{f,b}}\right].$$

$$(15)$$

Heat transfer rate is obtained as follows:

$$Q = \varepsilon C_{min}\left(T_{in,a} - T_{in,b}\right).$$

$$(16)$$

To simplify the computation, this study only considers the pressure drop of inlet channels, core part, and outlet channels. The total pressure drops for every side are defined as below:

$$\Delta P = \Delta P_1 - \Delta P_2 + \Delta P_3,$$

$$(17)$$

where Δp_1 is the pressure drop caused by the change of cross-sectional area from the deflector outlet to fin inlet. Δp_2 is the pressure drop caused by the change of cross-sectional area from fin inlet to deflector outlet. Δp_3 consists of two pressure drops, namely, frictional pressure drops and pressure drops, caused by change of channel area. They can be determined as follows:

$$\Delta P_1 = \frac{G^2}{2\rho_{in}} \left(1 - \alpha^2 + K_{in} \right),$$

$$\Delta P_2 = \frac{G^2}{2\rho_{out}} \left(1 - \alpha^2 - K_{out} \right),$$

$$\Delta P_3 = \frac{G^2}{2\rho_{in}} \left[2 \left(\frac{\rho_{in}}{\rho_{out}} - 1 \right) + \left(\frac{4 f L 1}{D} \right) \frac{\rho_{in}}{\overline{\rho}} \right],$$

(18)

where K_{in} and K_{out} are empirical coefficient and can be obtained from literature [21]. $\overline{\rho}$ is the average density of fluids, which is calculated by

$$\overline{\rho} = \frac{\rho_{in} + \rho_{out}}{2}.$$

(19)

Therefore, the pressure drop loss rate of fluids for the two sides is obtained by

$$\gamma_a = \frac{\Delta P_a}{P_{in,a}},$$

$$\gamma_b = \frac{\Delta P_b}{P_{in,b}}.$$

(20)

Dynamic Simulation Modeling of Intercooler

The dynamic time-delay of intercooler is one important factor that influences the maneuverability of marine gas turbine. Considering the operation characteristics of intercooler in practical application, lumped parameter model is used to construct the dynamic simulation model of intercooler. The following assumptions will be used for the analysis.

- The inside flow of fin channel is simplified as the one dimension and the velocity of fluids is uniform in the same section.
- The temperature of metal wall varies along with the direction of fluids flow and the radial temperature difference of metal wall is neglected.
- The heat capacity of gas is neglected compared with that of metal wall.

Considering the above assumptions, the mass conservation equations for the two sides can be given as

$$W_{in,a} = W_{out,a},$$

$$W_{in,b} = W_{out,b}. \tag{21}$$

And the energy conservation equations for the two sides are formulated as

$$V_a \frac{d}{dt}\left(\rho_{m,a} C_{p,m,a} T_{m,a}\right)$$

$$= W_{in,a} C_{p,in,a} T_{in,a} - W_{out,a} C_{p,out,a} T_{out,a}$$

$$- h_a A_{ef,a}\left(T_{m,a} - T_w\right),$$

$$V_b \frac{d}{dt}\left(\rho_{m,b} C_{p,m,b} T_{m,b}\right)$$

$$= W_{in,b} C_{p,in,b} T_{in,b} - W_{out,b} C_{p,out,b} T_{out,b}$$

$$+ h_b A_{ef,b}\left(T_w - T_{m,b}\right),$$

$$(22)$$

where $T_{m,a}$ and $T_{m,b}$ are the average temperature for the two sides and are calculated by

$$T_{m,a} = \frac{T_{in,a} + T_{out,a}}{2},$$

$$T_{m,b} = \frac{T_{in,b} + T_{out,b}}{2}.$$

$$(23)$$

The channel volumes for the two sides are obtained by

$$V_a = \frac{\left(H_a - t_{f,a}\right)\left(S_{f,a} - t_{f,a}\right) L1 N_a \left(L2 - 2\delta_s\right)}{S_{f,a}},$$

$$V_b = \frac{\left(H_b - t_{f,b}\right)\left(S_{f,b} - t_{f,b}\right) L1 N_b \left(L2 - 2\delta_s\right)}{S_{f,b}}.$$

$$(24)$$

The temperature of metal wall is obtained as follows:

$$M_w C_{p,w} \frac{d}{dt} T_w = h_a A_{ef,a} \left[T_{m,a} - T_w\right] - h_b A_{ef,b} \left[T_w - T_{m,b}\right].$$

$$(25)$$

In order to further raise the precision of simulation, the model considers related physical properties dependent on temperature.

DYNAMIC PERFORMANCE ANALYSIS AND DISCUSSIONS FOR INTERCOOLER

Based on effectiveness-number of transfer units (ε-NTU) and lumped parameter method mentioned earlier, the dynamic simulation model of intercooler is established by computer simulation software MATLAB/SIMULINK, which can be seen in Figure 4.

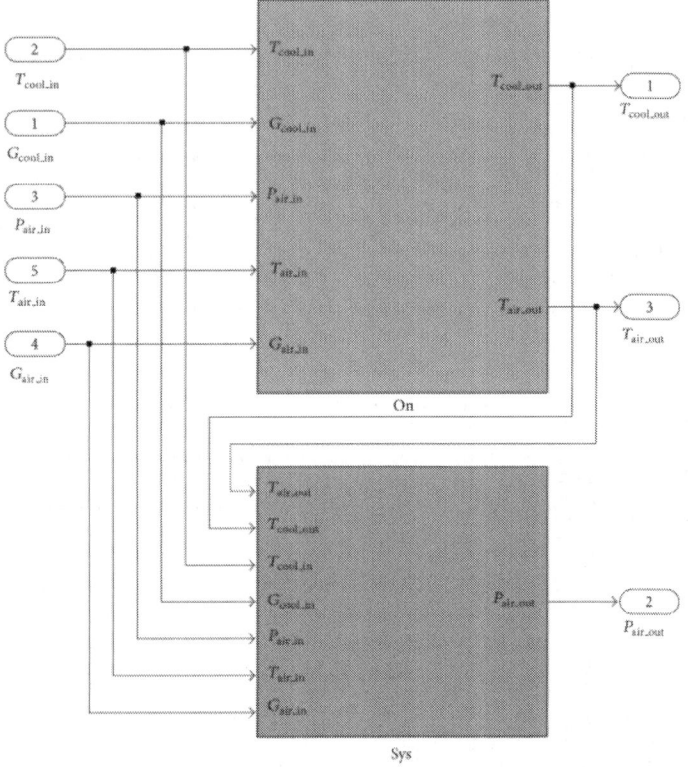

Figure 4: Dynamic simulation model of intercooler.

Model Validation

In order to verify the correctness and validity of the simulation model established in this paper, a case study taken from the work of Wen is considered [17]. The straight fin is used on the gas and coolant side. The preliminary structure of the intercooler is shown in Table 1. Table 2 lists the inlet parameters of gas under different gas turbine operation conditions. Water is chosen as the coolant and its inlet temperature and mass flow are assumed to be constant under different gas turbine operation conditions, which are 20°C and 200 kg/s, respectively. The intercooler is made of copper-nickel alloy. Some useful thermophysical parameters of copper-nickel alloy are mentioned in Table 3.

Table 1: Preliminary structure of the intercooler [17]

Parameters	Values
Number of gas side fin layers, Na	36
Fin pitch of gas side, Sf,a (m)	$1.4 \times 10{-3}$
Plate pitch of gas side, Ha (m)	$6.2 \times 10{-3}$
Fin thickness of gas side, tf,a (m)	$2 \times 10{-4}$
Number of coolant side fin layers, Nb	35
Fin pitch of coolant side, Sf,b (m)	$1.4 \times 10{-3}$
Plate pitch of coolant side, Hb (m)	$3 \times 10{-3}$
Fin thickness of coolant side, tf,b (m)	$2 \times 10{-4}$
Side plate thickness, dsp (m)	$1.6 \times 10{-3}$
Plate thickness, dp (m)	$5 \times 10{-4}$
Seal thickness, ds (m)	$6 \times 10{-3}$
Intercooler length, L1 (m)	0.35
Intercooler width, L2 (m)	0.4266

Table 2: The inlet parameters of gas under different gas turbine operation conditions [17]

Operation conditions	Inlet pressure (Pa)	Inlet temperature (°C)	Inlet mass flow (kg/s)
100%	302963	155.25	72.93
85%	288460	149.45	68.52
70%	277647	143.55	64.10
58%	258919	135.45	57.36
47%	240192	127.35	50.63

| 35% | 221464 | 119.25 | 43.89 |
| 17% | 170000 | 107.25 | 32.90 |

Table 3: Thermophysical parameters of different materials

Material type	Thermal conductivity (W/m·K)	Specific heat capacity (J/kg·K)	Density (kg/m3)
Copper-nickel alloy	38.5	380	8890
Copper	401	386	8960
Aluminum	237	897	2700

Figures 5, 6, and 7 show the comparisons of the simulation results in this paper with the thermodynamic calculation and numerical simulation results that have been reported in literature [17]. From Figures 5–7, it is obvious that the outlet temperature and pressure drop loss rate of gas and outlet temperature of coolant obtained by using the simulation model (variable properties) are basically consistent with the thermodynamic calculation results reported in literature [17]. The maximum deviation of pressure drop loss rate of gas is about 0.8% between simulation model (variable properties) and thermodynamic calculation method. In other words, the simulation model (variable properties) can be considered correct and feasible. The simulation results also show that both the outlet temperatures of gas and coolant can decrease when the operation conditions of gas turbine are reduced. In addition, the analysis and comparisons of the results demonstrate that the pressure drop loss rate of gas obtained by using the numerical simulation method is lower than that calculated by using other two methods at some operation conditions. The reason is that only the flow resistance loss of gas side is considered during the numerical simulation processes. Comparing the results obtained by two simulation models in this paper with those of literature [17], we also can conclude that the simulation model (variable properties) is more accurate and appropriate. Moreover, the convergence speed can be significantly improved when considering the influence of temperature on thermal physical property parameters. Therefore, the simulation model (variable properties) is used to analyze and discuss the dynamic performance of intercooler in the following section.

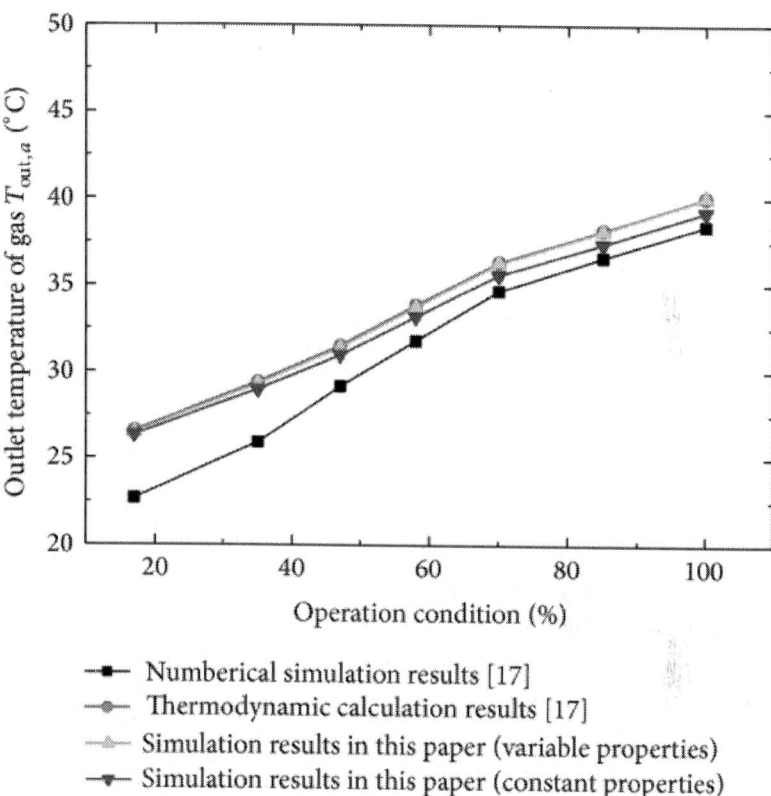

Figure 5: Comparisons of simulation outlet temperature of gas with the results reported in literature [17].

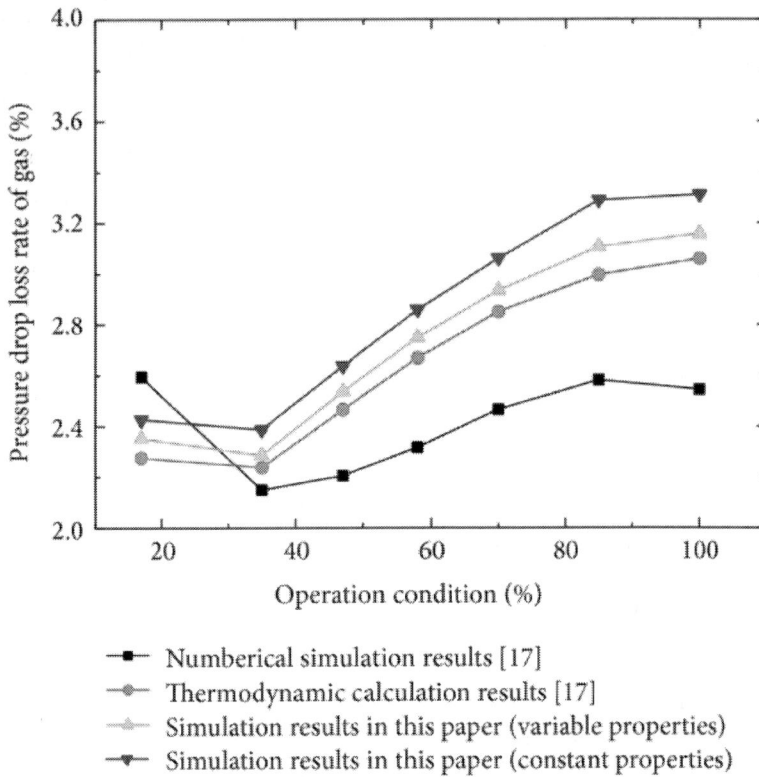

Figure 6: Comparisons of simulation pressure drop loss rate of gas with the results reported in literature [17].

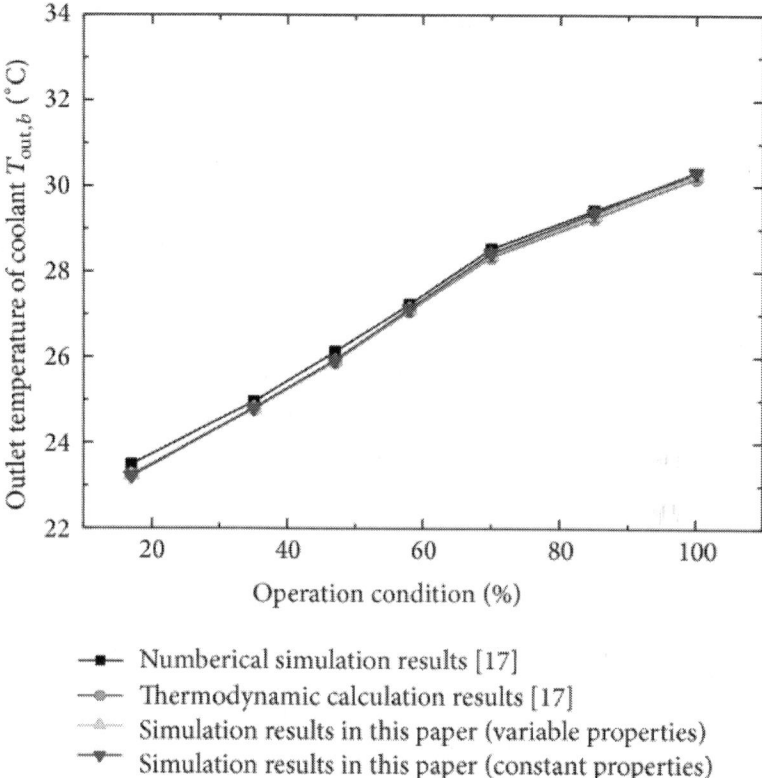

- —■— Numberical simulation results [17]
- —●— Thermodynamic calculation results [17]
- ⋯△⋯ Simulation results in this paper (variable properties)
- —▼— Simulation results in this paper (constant properties)

Figure 7: Comparisons of simulation outlet temperature of coolant with the results reported in literature [17].

Material Effects on Dynamic Performance

Material is one of the important factors which affect the structural strength and heat transfer performance of intercooler. In the following analysis, we will focus on investigating the effects of different material on dynamic performance of intercooler. The geometric dimensions of the intercooler and the thermophysical properties of materials (copper-nickel alloy, copper, and aluminum) studied here are listed in Tables 1 and 3. Water is chosen as the coolant and its inlet temperature and mass flow are assumed to be constant under different gas turbine operation conditions, which are 20°C and 200 kg/s, respectively. For the gas turbine, the operation condition linearly changes from 35% to 70% in 5 seconds and the related inlet operation parameters of gas are given in Table 2.

The dynamic response curve of the outlet temperature and pressure of gas and the outlet temperature of coolant with different materials in the change of operation condition of gas turbine are shown in Figures 8, 9, and 10. It may be clearly observed in Figures 8–10 that the outlet temperatures of gas and coolant change obviously in the previous stages and their gradients become smaller over time for all materials. The thermal inertia characteristic of intercooler is so obvious that it is necessary to consider the effect of intercooler on the thermodynamic performance of gas turbine. In addition, a careful inspection of Figures 8–10 reveals that intercooler made of copper or aluminum has the better heat transfer performance than that made of copper-nickel alloy. And the intercooler made of copper has the best heat transfer performance since it has the highest thermal conductivity. However, the dynamic response time of intercooler made of aluminum is the shortest as compared with the intercooler made of the other types of materials. This is due to the fact that aluminum has the lowest density. The larger the value of density is, the longer the dynamic response time of the intercooler will be. From Table 3, it can be seen that the three materials, ordered by decreasing density, are (1) copper, (2) copper-nickel alloy, and (3) aluminum. These suggest that the weight of intercooler is the important factor on affecting the thermal inertia of intercooler, which is consistent with the result of the other literatures [16]. Additionally, the outlet pressure of gas is less affected by materials and there is no obvious flow time delay characteristic because gas has a higher velocity.

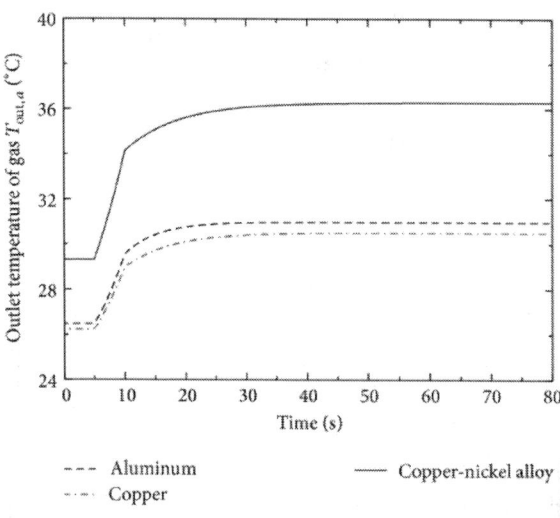

Figure 8: Dynamic response curve of the outlet temperature of gas with different materials.

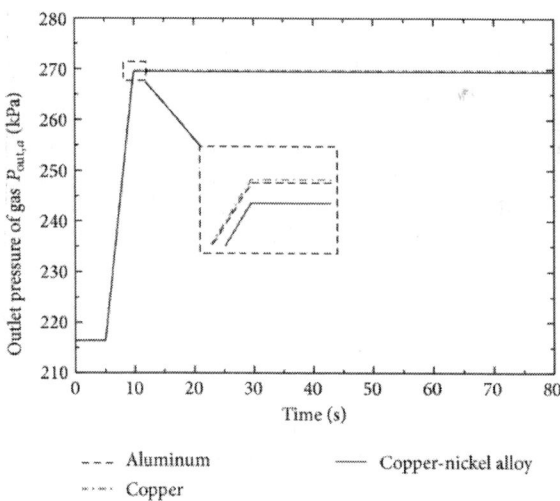

Figure 9: Dynamic response curve of the outlet pressure of gas with different materials.

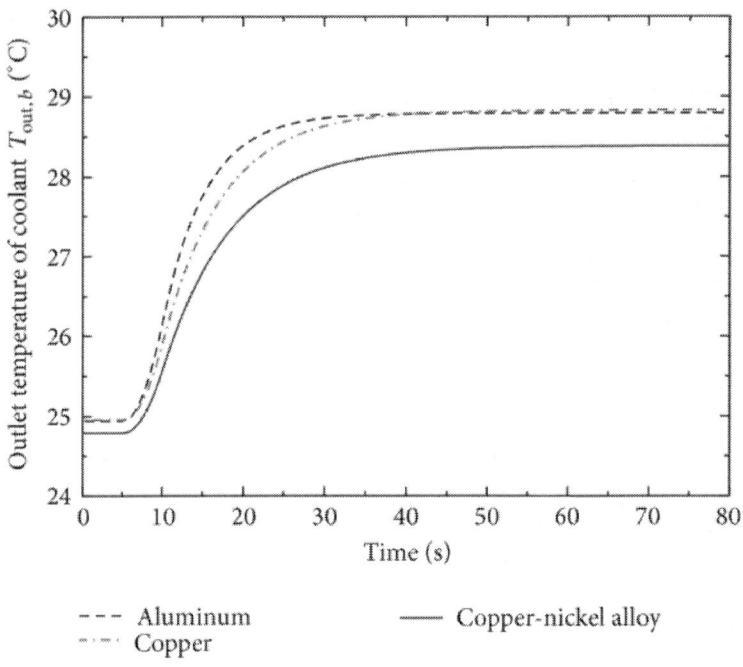

Figure 10: Dynamic response curve of the outlet temperature of coolant with different materials.

To sum up, material is a significant factor which affects the heat transfer and dynamic performance of intercooler. From the point of view of heat transfer performance and dynamic time-delay characteristics, aluminum has a slightly better performance than the other materials. However, considering the application environment, copper-nickel alloy is often used as the material of intercooler because it has good anticorrosion and excellent strength.

Coolant Effects on Dynamic Performance

Although the copper-nickel alloy has good anticorrosion and excellent strength, its heat transfer performance is relatively poor and dynamic delay response time is longer than the other materials. Therefore, it is very important to improve the heat transfer and dynamic performance of intercooler from other aspects.

Coolant seems to be another crucial factor in determining the heat transfer and dynamic performance of intercooler. As the most common and cheap coolant, water freezes at a low temperature, which can lead to intercooler failure. In order to solve this problem, ethylene glycol mixed with water in different volume percentages is typically used to lower the aqueous freezing point of the heat transfer medium in the practical industrial applications [22]. Ethylene glycol and water (EG/water) can withstand low temperatures down to $-60°C$ [23]. However, it can cause an erosive action on the intercooler, which causes fouling and affects the heat transfer performance of intercooler. Besides, this fluid mixture is toxic so that it is a potential danger for staff.

In the following analysis, we are interested in investigating the effects of different coolants on dynamic performance of intercooler. Water and EG/water (50 : 50 by mass) are chosen as the coolant and the inlet temperature and mass flow of two coolants are assumed to be constant under different gas turbine operation conditions, which are 20°C and 200 kg/s, respectively. The properties of two coolants at different temperature are available in ASHRAE [24]. The intercooler is made of copper-nickel alloy. For the gas turbine, the operation condition linearly changes from 35% to 70% in 5 seconds and the related inlet operation parameters of gas are given in Table 2.

Figures 11, 12, and 13 show the dynamic response curve of the outlet temperature and pressure of gas and the outlet temperature of coolant with different coolants in the change of operation condition of gas turbine. From Figures 11–13, it is easy to see that the intercooler has better heat transfer performance and smaller thermal inertia when water is used as coolant. The outlet temperature and dynamic delay response time of gas obtained by using water as coolant can reduce about 3°C and 15 seconds, respectively, compared with those obtained by using EG/water as coolant. The main reason may come from the fact that water has higher heat capacity and thermal conductivity and lower viscosity than EG/water. In other words, the coolants with higher thermal conductivity and heat capacity but lower viscosity show better heat transfer performance and thermal inertia. Meanwhile, the results also show that the dynamic delay response time of gas is shorter than that of every coolant. This is due to the fact that coolants have higher heat capacity and thermal conductivity. Additionally, the outlet pressure of gas is less affected by coolants and there is no obvious flow time delay characteristic.

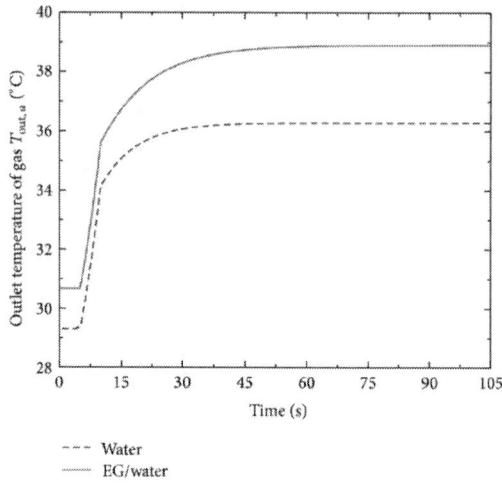

Figure 11: Dynamic response curve of the outlet temperature of gas with different coolants.

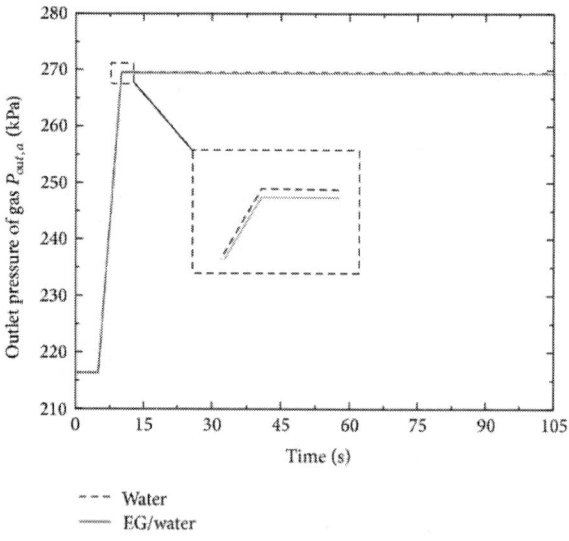

Figure 12: Dynamic response curve of the outlet pressure of gas with different materials.

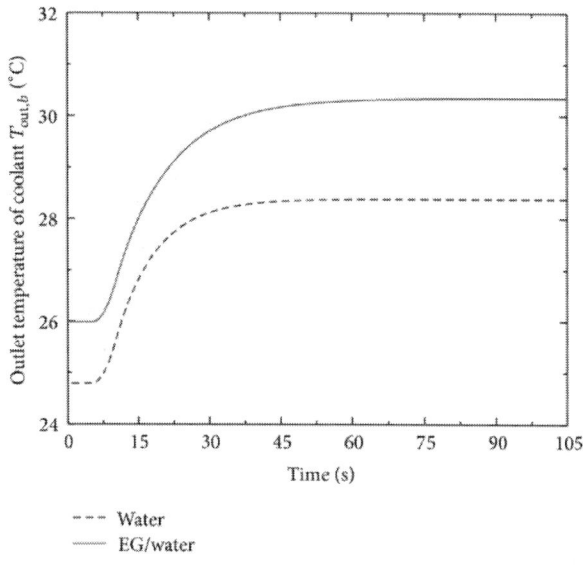

Figure 13: Dynamic response curve of the outlet temperature of coolant with different materials.

These findings suggest that all the applied conditions of coolants, corrosion resistant properties of materials, and the flow heat transfer performance requirements of intercooler should be considered when selecting the coolants for intercooler.

OPTIMAL DESIGN OF INTERCOOLER BASED ON SIMULATED ANNEALING ALGORITHM

The design of intercooler involves a large number of geometric and operating variables that need to meet the flow and heat transfer performance requirement under some constraints [15]. The conventional optimization methods become very cumbersome and laborious to solve the optimization problem. In recent times, some nontraditional probabilistic search algorithms, namely, genetic algorithm (GA) [25–27], particle swarm optimization (PSO) algorithm [28], and harmony search (HS) algorithm [20], have been applied to the optimization of various heat exchangers. Wang et al. [25] presented the GA for the optimization of microturbine recuperators from technical and economic standpoints and discussed the solution strategies under two different fitness functions. Their results showed that GA has good global search capability to realize the compact design of recuperators. However, GA is not very effective for the local search space problem. Compared with GA, PSO algorithm has the ability of memory. But PSO algorithm has a shortcoming of converging prematurely after getting trapped into some local optima and considers it to be the global optima. Besides, the acceleration constants and inertia weight should be given reasonably since they are employed to control the exploration abilities of the swarm and affect the convergence behavior. When PSO algorithm is applied to a multidimensional complex problem scenario, it becomes nearly impossible to get out from that local optima and reach out for the global optima due to some constraints. Moreover, the evolutionary algorithms, such as GA, PSO algorithm, and HS algorithm, cannot deal with the constraints directly and many constraint handling methods should be employed to help the optimization process [20].

As previously mentioned, it is a constraints optimization problem for the structure optimization design of intercooler. The penalty function method is often used to transfer constrained condition into unconstrained condition [20, 26], which may affect the optimization results. As a stochastic optimization technique, simulated annealing algorithm was first put forward by Metropolis in 1953, and it was employed to seek an optimal combination by Kirkpatrick et al. until 1983 [29]. This algorithm simulates the thermodynamic process of slow cooling of molten metals to achieve the minimum function value in an optimization problem, so it has been widely used in solving sophisticated optimization problems [30]. Compared with GA and PSO algorithm, SA algorithm can hinder the premature convergence to the local optima and diverge the particles using its strong ability of local search.

Simulated Annealing Algorithm

For the following optimization problem, the basic optimization process based on simulated annealing algorithm is shown in Figure 14:

$$\min \quad f(X) \quad X = [x_1, x_2, \dots, x_n]$$
$$\text{s.t.} \quad \begin{cases} g_i(X) \le 0, & i = 1, 2, \dots, m, \\ h_j(X) = 0, & j = 1, 2, \dots, l, \end{cases}$$

$$(26)$$

where $f(x)$ is the objective function. $g_i(x)$ and $h_j(x)$ are the constraint equations and equilibrium equations, respectively.

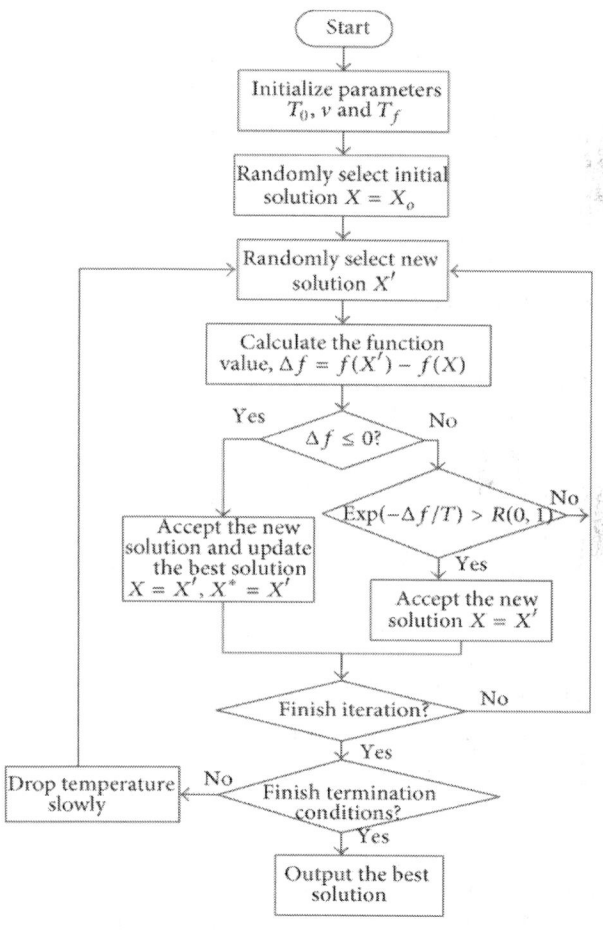

Figure 14: Flowchart of simulated annealing algorithm applied for the optimization of intercooler.

From Figure 14, we can see that simulated annealing algorithm mainly includes two circle processes, inner and outer circle processes. The purpose of the outer circle processes is to decrease annealing temperature. In every iteration of inner circle processes, the new solution is obtained and evaluated by the Metropolis criterion to determine whether the new solution will be accepted or not. The following shows the details of the basic optimization processes.

Step 1: Set initial temperature $T=T_0 (T_0 > 0)$, cooling rate v, and final temperature T_f.

Step 2: Randomly select initial solution X and the approximation of the optimal solution X* from all possible solutions, $X=X^*=X_0$.

Step 3: Randomly select disturbance to obtain the new solution X' from the sets of all possible neighbors of X.

Step 4: Calculate the function value f(X) and f(X') by objective function, respectively. Df = f(X') − f(X) .

Step 5: If Df ≤ 0, then the new solution is accepted and the approximation of the optimal solution is updated, $X=X', X^*=X'$.

Step 6: If Df > 0, randomly select R from uniform distribution on the interval (0,1). If p(Df)=exp(-Df/T) > R(0,1), then the new solution is accepted but it is a worse solution, $X=X'$. Or else, the current solution remains unchanged.

Step 7: Repeat the above Steps 3–6, until loop iteration steps meet the requirements.

Step 8: Check termination criterion and output the optimal solution.

Simulated annealing algorithm uses the cooling process and the Metropolis algorithm to control the search process, so this algorithm can leap from the local minimum during the search and handle any type of variable easily, including noncontinuous functions and nondifferential functions [29].

Objective Function, Optimization Variables, and Constraints

The results reported in literature [16] showed that the weight of intercooler is an important factor to affect its thermal inertia. With the increase of the weight of intercooler, the thermal inertia increases rapidly. Therefore, the total weight of intercooler is selected as the optimal objective function in this study:

$$\min M_w = \rho_w \left(V - V_a - V_b \right).$$

(27)

When putting all the relevant values, the above equation can be simplified and expressed as

$$\min M_w = \rho_w \left[L1L2 \left(N_a H_a + N_b \left(H_b + 2\delta_p \right) + 2\delta_{sp} \right) \right.$$

$$- \frac{\left(H_a - t_{f,a} \right) \left(S_{f,a} - t_{f,a} \right) L1 N_a \left(L2 - 2\delta_s \right)}{S_{f,a}}$$

$$\left. - \frac{\left(H_b - t_{f,b} \right) \left(S_{f,b} - t_{f,b} \right) L1 N_b \left(L2 - 2\delta_s \right)}{S_{f,b}} \right].$$

(28)

Considering the basic assumptions mentioned above, the plate pitches for the two sides, fin pitch, fin thickness, intercooler length, intercooler width, and numbers of fin layers for gas side are taken as the optimization variables in this study. All the above optimization variables are considered as structural constraints, which are expressed as

$$\text{s.t.} \begin{cases} g_1(X) \Longrightarrow H_{a,\min} \leq H_a \leq H_{a,\max}, \\ g_2(X) \Longrightarrow H_{b,\min} \leq H_b \leq H_{b,\max}, \\ g_3(X) \Longrightarrow S_{f,a,\min} \leq S_{f,a} \leq S_{f,a,\max}, \\ g_4(X) \Longrightarrow t_{f,a,\min} \leq t_{f,a} \leq t_{f,a,\max}, \\ g_5(X) \Longrightarrow L1_{\min} \leq L1 \leq L1_{\max}, \\ g_6(X) \Longrightarrow L2_{\min} \leq L2 \leq L2_{\max}, \\ g_7(X) \Longrightarrow N_{a,\min} \leq N_a \leq N_{a,\max}. \end{cases}$$

(29)

In order to achieve the design requirements, the heat transfer efficiency required of intercooler and the pressure drop loss rate of gas are considered as the performance constraints, which can be given as follows:

$$\text{s.t.} \begin{cases} g_8\left(X\right) \Longrightarrow \varepsilon \geq \varepsilon_{\min}, \\ g_9\left(X\right) \Longrightarrow \gamma_a \leq \gamma_{a,\max}. \end{cases}$$

$$(30)$$

Optimal Design Results of Intercooler

This section explores the use of simulated annealing (SA) algorithm for structure optimization of intercooler and discusses the effects of structure optimization on dynamic performance of intercooler. The intercooler is made of copper-nickel alloy. Water is chosen as the coolant and the inlet temperature and mass flow of two coolants are assumed to be constant under different gas turbine operation conditions, which are 20°C and 200 kg/s, respectively. Gas turbine operates in a 100% condition.

In this study, the seven design variables such as the plate pitches for the two sides, fin pitch, fin thickness, intercooler length, intercooler width, and numbers of fin layers for gas side are selected as the optimization variables. All variables are continuous except the number of gas side layers. The other structural parameters of intercooler such as plate thickness, seal thickness, and side plate thickness are considered to be constant as listed in Table 1 and are not to be optimized. The variation ranges of the design variables and performance constraints of intercooler have been mentioned in Table 4.

Table 4: Variation ranges of design parameters and performance constraints of intercooler

Parameters	Minimum	Maximum
Number of gas side fin layers, Na	30	40
Fin pitch of gas side, Sf,a (m)	1.5×10^{-3}	3×10^{-3}
Plate pitch of gas side, Ha (m)	4×10^{-3}	7×10^{-3}

Fin thickness of gas side, tf,a (m)	$1.5 \times 10{-}4$	$3 \times 10{-}4$
Plate pitch of coolant side, Hb (m)	$2 \times 10{-}3$	$4.5 \times 10{-}3$
Intercooler length, L1 (m)	0.31	0.38
Intercooler width, L2 (m)	0.400	0.500
Effectiveness, ε (%)	85	
Pressure drop loss rate of gas side, ga (%)		5

Figure 15 shows the effectiveness convergence diagram as objective function. A significant decrease in the objective function has been observed in the beginning of the evaluation process after 100 iterations. After approximate 6000 iterations, the changes in the objective function become relatively low. Finally, the minimum weight of intercooler is found after 7000 iterations with the value of 1139.87 kg. Tables 5 and 6 show the preliminary design of intercooler and optimum structure which are obtained from simulated annealing algorithm. The increment of 2.07% has been observed in efficiency of intercooler by optimization method in comparison with preliminary design. Besides, the weight of intercooler has decreased from 1391.45 kg to 1139.87 kg while the pressure drop loss rate of gas side has increased from 3.17% to 3.62%.

Table 5: The structure comparisons of preliminary design and simulated annealing algorithm

Parameters	Preliminary design	Optimal design
Number of gas side fin layers, Na	36	37
Fin pitch of gas side, Sf,a (m)	$1.4 \times 10{-}3$	$1.1 \times 10{-}3$
Plate pitch of gas side, Ha (m)	$6.2 \times 10{-}3$	$6.3 \times 10{-}3$
Fin thickness of gas side, tf,a (m)	$2 \times 10{-}4$	$1.5 \times 10{-}4$
Plate pitch of coolant side, Hb (m)	$3 \times 10{-}3$	$2.1 \times 10{-}3$
Intercooler length, L1 (m)	0.350	0.3222
Intercooler width, L2 (m)	0.4266	0.4045

Table 6: The performance comparisons of preliminary design and simulated annealing algorithm

Parameters	Preliminary design	Optimal design
Effectiveness, ε (%)	84.97	87.04
Pressure drop loss rate of gas side, ga (%)	3.17	3.62
Weight, Mw (kg)	1391.45	1139.87

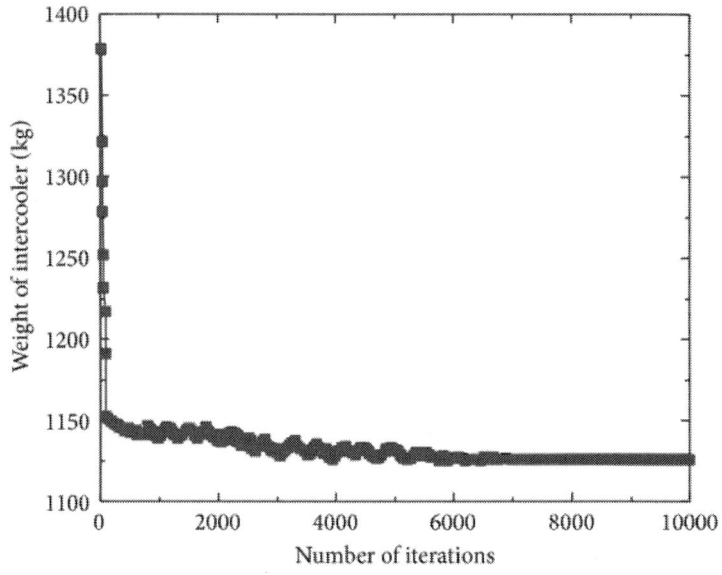

Figure 15: Evolution process for minimum weight based on simulated annealing algorithm.

The dynamic response curve of the outlet temperature and pressure of gas and outlet temperature of coolant with different intercoolers (preliminary design and optimum structure) in the linear change of operation condition of gas turbine from 35% to 70% in 5 seconds are shown in Figures 16, 17, and 18. From Figures 16–18, it is clear that the intercooler optimized by simulated annealing algorithm has better heat transfer performance and smaller thermal inertia. The outlet temperature of gas obtained by simulated annealing algorithm can reduce about 3°C compared with that obtained by preliminary design. The dynamic delay response times of gas and coolant sides of preliminary intercooler are 11.5 seconds and 20 seconds longer than those obtained by the simulated annealing algorithm. These findings suggest that optimal design based on artificial intelligence algorithm is available and necessary to realize the high compact design of intercooler, which can further affect the overall performance of intercooled cycle marine gas turbine.

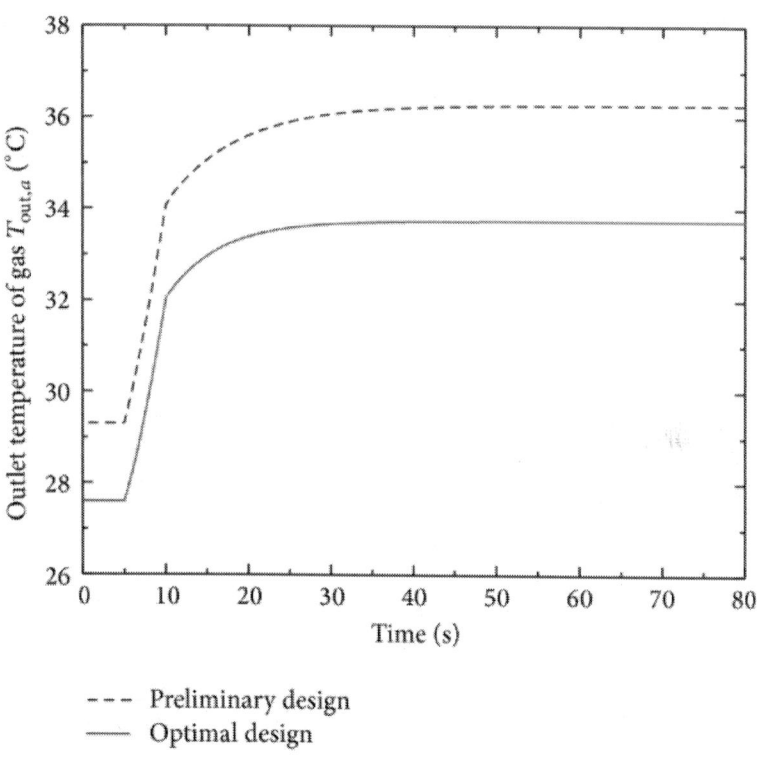

- - - Preliminary design
——— Optimal design

Figure 16: Dynamic response curve of the outlet temperature of gas with different intercoolers.

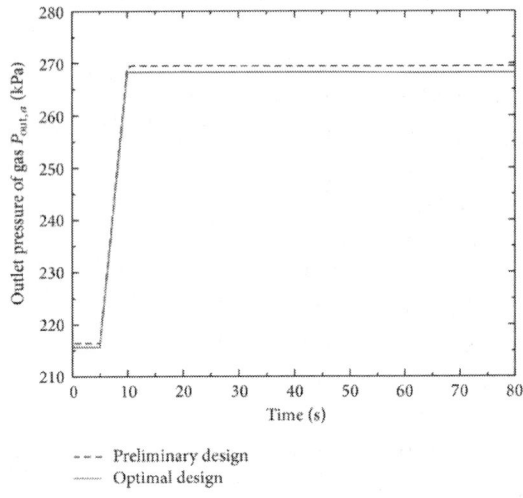

Figure 17: Dynamic response curve of the outlet pressure of gas with different intercoolers.

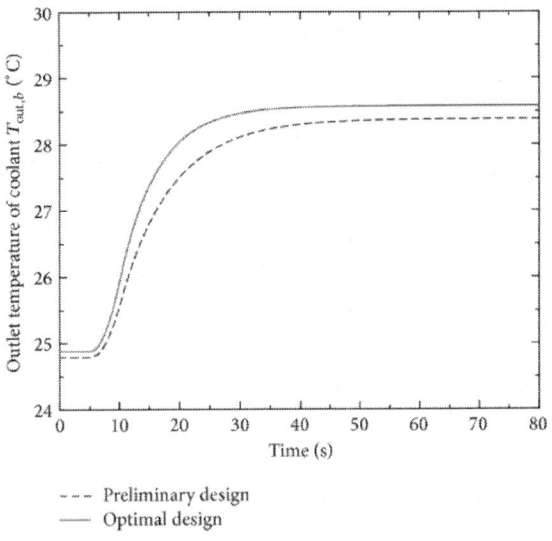

Figure 18: Dynamic response curve of the outlet temperature of coolant with different intercoolers.

CONCLUSIONS

This study studied the dynamic time-delay characteristics of marine gas turbine intercooler and showed the successful usage of the simulated annealing (SA) algorithm in the optimal design of intercooler. On the basis of the work presented in this study, the following conclusions could be made. A generalized simulation model was developed to carry out the dynamic performance analysis of intercooler based on the effectiveness-number of transfer units and lumped parameter method. The analytical results of an example showed that the simulation model (variable properties) established in this study was correct and feasible. The effects of substrate materials and coolants on the dynamic time-delay characteristics of intercooler were analyzed in detail. The results showed that both material and coolant were the significant factors that affected the heat transfer and dynamic performance of intercooler. For all materials and coolants, gas side had a slightly smaller thermal inertia than liquid side. When other conditions were constant, better thermal performance and smaller thermal inertia were noted for the materials with higher thermal conductivity and lower density than that for the material with lower thermal conductivity and higher density. Besides, the coolant with higher thermal conductivity and specific heat but lower viscosity was beneficial to improve the heat transfer performance and thermal conductivity of intercooler. However, the outlet pressure of gas was less affected by materials and coolants, and there was no obvious flow time-delay characteristic. The present study demonstrated successful application of simulated annealing technique for the structure optimization of intercooler considering minimum weight as objective functions. The improvement of heat transfer and dynamic performance was observed for intercooler obtained by using simulated annealing algorithm compared with preliminarily design, showing the improvement potential of the artificial intelligent technique for intercooler optimization.

REFERENCES

1.   X. Y. Wen and D. M. Xiao, "Feasibility study of an intercooled-cycle marine gas turbine," Journal of Engineering for Gas Turbines and Power, vol. 130, no. 2, Article ID 022201, 2008. ··

2.   X. Y. Wen and D. M. Xiao, "A new concept concerning the development of high-power marine gas turbines," Ship Science and Technology, vol. 29, no. 4, pp. 17–21, 2007 (Chinese).

3.   S. B. Shepard, T. L. Bowen, and J. M. Chiprich, "Design and development of the WR-21 intercooled recuperated (ICR) marine gas turbine," Journal of Engineering for Gas Turbines and Power, vol. 117, no. 3, pp. 557–562, 1995.

4.   S. Y. Li, Z. T. Wang, J. Q. Wang, and P. P. Luo, "Simulation study on fuel supply rate curve of marine inter-cooled gas turbine," Ship Engineering, vol. 32, no. 5, pp. 15–18, 2010 (Chinese).

5.   Y. L. Ying, Y. P. Cao, and S. Y. Li, "Research on fuel supply rate of marine intercooled-cycle engine based on simulation experiment," International Journal of Computer Applications in Technology, vol. 47, no. 4, pp. 212–221, 2013.

6.   Y. L. Ying, Y. P. Cao, S. Y. Li, and Z. T. Wang, "Study on flow parameters optimisation for marine gas turbine intercooler system based on simulation experiment," International Journal of Computer Applications in Technology, vol. 47, no. 1, pp. 56–67, 2013. · ·

7.   W. H. Wang, L. G. Chen, and F. R. Sun, "Power optimization of a real closed intercooled regenerated gas turbine cycle," Chinese Journal of Mechanical Engineering, vol. 41, no. 4, pp. 55–58, 2005.

8.   W. H. Wang, L. G. Chen, F. R. Sun, and C. Wu, "Performance optimisation of open cycle intercooled gas turbine power plant with pressure drop irreversibilities," Journal of the Energy Institute, vol. 81, no. 1, pp. 31–37, 2008. · ·

9.   W. H. Wang, L. G. Chen, and F. R. Sun, "Thermodynamic optimization of a triple-shaft open intercooled, recuperated gas turbine cycle. Part 1: description and modeling," International Journal of Low-Carbon Technologies, 2013. ·

10. W. H. Wang, L. G. Chen, and F. R. Sun, "Thermodynamic optimization of a triple-shaft open intercooled, recuperated gas turbine cycle. Part 2: power and efficiency optimization," International Journal of Low-Carbon Technologies, 2013.

11. C. Z. Wen and W. Dong, "Numerical simulation of heat transfer and fluid flow on marine gas turbine intercooler," Journal of Aerospace Power, vol. 25, no. 3, pp. 654–658, 2010.

12. W. Dong, C. Mao, J. J. Zhu, and Y. Chen, "Numerical and experimental analysis of inlet non-uniformity influence on intercooler performance," in Proceeding of the ASME Turbo Expo 2012: Turbine Technical Conference and Exposition (GT ‹12), pp. 349–357, Copenhagen, Denmark, June 2012. ··

13. P. Gao and W. Dong, "Optimal analysis of flow parameters for marine gas turbine intercooler,"Aeroengine, vol. 37, no. 4, pp. 29–32, 2011 (Chinese).

14. Z. Li, H. B. Zhang, X. Y. Wen, and D. M. Xiao, "Numerical simulation of an intercooler for a complex-cycle Marine gas turbine," Journal of Engineering for Thermal Energy and Power, vol. 23, no. 2, pp. 148–152, 2008 (Chinese).

15. X. Xiao, The optimization design, modeling and control of the gas turbine intercooler [M.S. thesis], Shanghai Jiao Tong University, 2013, (Chinese).

16. S. K. Zhang, Simulation research on performance of marine intercooled cycle gas turbine [M.S. thesis], China Ship Research and Development Academy, 2012, (Chinese).

17. C. Z. Wen, Design and study on intercooling heat exchanger of marine gas turbine [M.S. thesis], Shanghai Jiao Tong University, 2009, (Chinese).

18. V. Gnielinski, "New equations for heat and mass transfer in turbulent pipe and channel flow (Neue Gleichungen für den Wärme—und den Stoffübergang in turbulent durchströmten Rohren und Kanälen)," Forschung im Ingenieurwesen, vol. 41, no. 1, pp. 8–16, 1975.

19. H. Peng and X. Ling, "Optimal design approach for the plate-fin heat exchangers using neural networks cooperated with genetic algorithms," Applied Thermal Engineering, vol. 28, no. 5-6, pp. 642–650, 2008. ··

20. M. Yousefi, R. Enayatifar, A. N. Darus, and A. H. Abdullah, "Optimization of plate-fin heat exchangers by an improved harmony search algorithm," Applied Thermal Engineering, vol. 50, no. 1, pp. 877–885, 2013. ··

21. S. H. Wang, Plate-Fin Heat Exchanger, Chemical Industry Press, Beijing, China, 1984 (Chinese).

22. F. C. McQuiston, J. D. Parker, and J. D. Spitler, Heating Ventilating

and Air Conditioning, John Wiley & Sons, New York, NY, USA, 2000.

23.   P. K. Namburu, D. K. Das, K. M. Tanguturi, and R. S. Vajjha, "Numerical study of turbulent flow and heat transfer characteristics of nanofluids considering variable properties," International Journal of Thermal Sciences, vol. 48, no. 2, pp. 290–302, 2009. · ·

24.   ASHRAE Handbook Fundamentals, American Society of Heating, Refrigerating and Air-Conditioning Engineers, Atlanta, Ga, USA, 2005.

25.   Q. W. Wang, H. X. Liang, G. N. Xie, M. Zeng, L. Q. Luo, and Z. P. Feng, "Genetic algorithm optimization for primary surfaces recuperator of microturbine," Journal of Engineering for Gas Turbines and Power, vol. 129, no. 2, pp. 436–442, 2007. · ·

26.   G. N. Xie, B. Sunden, and Q. W. Wang, "Optimization of compact heat exchangers by a genetic algorithm," Applied Thermal Engineering, vol. 28, no. 8-9, pp. 895–906, 2008. · ·

27.   L. Gosselin, M. Tye-Gingras, and F. Mathieu-Potvin, "Review of utilization of genetic algorithms in heat transfer problems," International Journal of Heat and Mass Transfer, vol. 52, no. 9-10, pp. 2169–2188, 2009. · · ·

28.   R. V. Rao and V. K. Patel, "Thermodynamic optimization of cross flow plate-fin heat exchanger using a particle swarm optimization algorithm," International Journal of Thermal Sciences, vol. 49, no. 9, pp. 1712–1721, 2010. · ·

29.   S. Kirkpatrick, J. Gelatt, and M. P. Vecchi, "Optimization by simulated annealing," American Association for the Advancement of Science: Science, vol. 220, no. 4598, pp. 671–680, 1983. · · ·

30.   J. M. Reneaume and N. Niclout, "Plate fin heat exchanger design using simulated annealing," Computer Aided Chemical Engineering, vol. 9, pp. 481–486, 2001. · ·

Chapter 8

Gas Turbine Engine Control Design Using Fuzzy Logic and Neural Networks

M. Bazazzadeh, H. Badihi, and A. Shahriari

Department of Mechanical & Aerospace Engineering, Malek-Ashtar University of Technology, Shahin Shahr, Isfahan, Isf, 83145/115, Iran

ABSTRACT

This paper presents a successful approach in designing a Fuzzy Logic Controller (FLC) for a specific Jet Engine. At first, a suitable mathematical model for the jet engine is presented by the aid of SIMULINK. Then by applying different reasonable fuel flow functions via the engine model, some important engine-transient operation parameters (such as thrust, compressor surge margin, turbine inlet temperature, etc.) are obtained. These parameters provide a precious database, which train a neural network. At the second step, by designing and training a feedforward multilayer perceptron neural network according to this available database; a number of different reasonable fuel flow functions for various engine acceleration operations are determined. These functions are used to define the desired fuzzy fuel functions. Indeed, the neural networks are used as an effective method to define the optimum fuzzy fuel functions. At the next step, we propose a FLC by using the engine simulation model and the neural network results. The proposed control scheme is proved by computer simulation using the designed engine model. The simulation results of engine model with FLC illustrate that the proposed controller achieves the desired performance and stability.

INTRODUCTION

Gas turbine engines are constituted of a complex system. Their desired performance can guarantee the aircraft flight safety. This performance is impressed by some engine input controlling functions which are changing with the development of engines. Finding these functions can be a great success in jet engine control issue [1, 2]. For example, when an airplane loses one of its operative engines during takeoff, the remaining engines should achieve their maximum thrust as fast as possible without any spools over speed, turbine inlet over temperature, and compressor surge. On the other hand, at a complex maneuvering by a fighter airplane, the engines should deliver more accurate thrust functions. Both of these examples can indicate the importance of the engine control issue.

While the traditional control techniques for aero-engines are time-tested and reliable, modern control techniques promise to provide improved control and therefore improved aircraft propulsion system performance [3, 4].

During the last five decades, most of control problems have been formulated by the objective knowledge of the given systems (e.g., mathematical model). Many of these "model-base" approaches have found their way into practice and provided satisfactory solutions to the spectrum of complex systems [5].

However, there are many knowledge-based systems that cannot be merely described by the traditional mathematical representations [6, 7].

The knowledge-based approaches are much closer to human thinking than traditional classical approaches [6,8].

In a broad perspective, knowledge-based approaches underlie what is called "soft computing". These methods include fuzzy logic (FL), neural networks (NN), genetic algorithms (GA), and probabilistic reasoning (PR). In addition, these methodologies in most part are complimentary rather than competitive [9].

Fuzzy logic has been the area of heated debate and much controversy during the last decades. The first paper in fuzzy set theory, which is now considered to be the seminar paper of the subject, was written by Zadeh (See [10]), who is considered the founding father of the field. In that work, Zadeh was implicitly advancing the concept of approximate human reasoning to make effective decisions on the basis of the available imprecise, linguistic information. The first implementation of Zadeh's idea was accomplished by Mamdani (see [11]) which demonstrated the ability of Fuzzy Logic Controller (FLC) for a small model steam engine. After this pioneer work, many consumer products and industrial applications using fuzzy technology have been developed and are currently available in whole of the world.

Fuzzy logic is one of the most effective approaches for intelligent control of complex nonlinear systems like aero-engines [4]. One of the important advantages of this approach is the simplicity of utilization. The other one is feasibility of increasing the number and type of membership functions and rules while it has wide variety of rules definition.

The engines of today are reliable and safe, but they are expensive to operate and maintain. With the integration of available models and algorithms for on-board operation and the resultant increase in on-board intelligence, engines of the future will be able to operate even more safely and reliably with reduced life cycle cost [3, 4].

With a renewed emphasis on reducing engine life cycle costs, improving fuel efficiency, increasing durability and life, and so forth, driven by various development programs, there is a strong push to do research on application of intelligent technologies for the engines.

The intelligent controlling approaches will provide the required scope for aero-engines to be more efficient, safe, and economic. These approaches offer the potential for creating extremely safe, highly reliable systems. The approaches will help to enable a level of performance that far exceeds that of today's propulsion systems in terms of reduction of harmful emissions, maximization of fuel efficiency, and minimization of noise, while improving system affordability and safety.

During the last decade, some research works have been carried out for intelligent control of gas turbine engines using fuzzy logic. In [12], the authors studied a combination of two potential techniques; fuzzy logic and evolutionary algorithms for a specific gas turbine. The controlling parameters include Inlet Guide Vane (IGV) angle and nozzle area. Similar work which has been presented in [13] is designing and evaluating two types of FLCs by controlling the combustor pressure. However, in the foresaid works and other related studies (see [14,15]), on the one hand, a comprehensive FLC has not been proposed that takes the flame-out and the safety limitations into account. Or on the other hand, the effect of fuel flow rate is not considered as a remarkable controlling parameter.

Generally, designing an appropriate controller for each specific aero-engine includes two steps: first, an accurate aero-engine model to simulate the engine behavior individually and second, system recognition. The system recognition as the most important step in creating suitable membership functions and rules is the knowledge of engine performance parameters and specially their relation during engine operation. One of the most important performance parameters is the fuel flow rate and its relation with turbine inlet temperature and compressor surge margin during various engine acceleration maneuverings.

In this paper, after designing an accurate simulator model for the engine, we consider the fuel flow rate directly as the most comprehensive controlling parameter which has significant effects on all engine performance parameters. A successful approach for achieving the system recognition is also presented. The proposed approach includes designing and training a neural network which simulate the engine performance backwardly. So it empowered us to obtain a series of fuel flow functions for various engine acceleration maneuverings. These functions were used in the process of membership functions definition.

TURBINE ENGINE SIMULATOR MODEL (TESM)

Engine performance characteristics at design point are important parameters obtained from engine design process. For initial definition work, the operating condition where an engine will spend most time has been traditionally chosen as the engine design point. Alternatively some important high power condition may be chosen. Either way, at the design point the engine configuration, component design and cycle parameters are optimised. A change to the engine design requires a different engine geometry, at a fixed operating condition [16].

The objective here is not to design an engine. Hence, there is a specific engine with fixed geometry and known parameters. We only seek to simulate different operating conditions of the engine including steady and transient operations.

There are several ways to do engine modeling and simulation such as traditional approaches, FORTRAN, or other 4th generation languages to calculate thermodynamics cycle parameters associated with the design and performance prediction of gas turbine engines. These cycle decks have been used extensively by developers of gas turbine engines to understand the behavior of jet engine designs prior to, during, and after the development of the physical engines. In addition to them, Simulink software and associated tools can be applied in the development of detailed, physics-based, turbine engine models.

Simulink provides an easy-to-use, graphical, modeling and simulation development environment for developing time-based simulations in a wide range of applications, and Simulink is capable of code generation using associated tools [17–19]. So, a dynamic model for a specific turbojet engine is developed using the MATLAB simulation environment and its Simulink toolbox [1]. The schematic configuration of the simulated turbojet engine is shown in Figure 1.

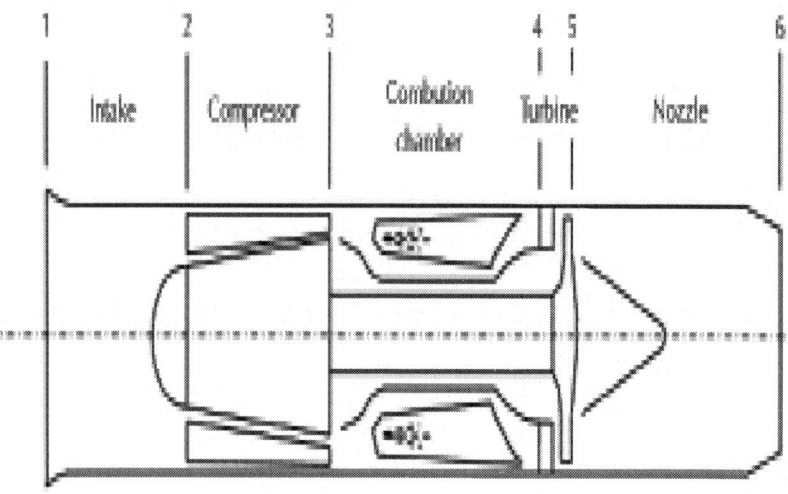

Figure 1: Schematic configuration of the simulated turbojet engine.

The engine model is constructed with a component approach for ease of modification and replacement with different engine components. The precision of Turbine Engine Simulator Model (TESM) results depends on the precision of each component module result [1].

Each component can be instantiated from a software library module developed to represent the functions of that particular type of component. Each module is a functional unit with its own set of inputs and outputs (I/O). Each can function as an independent component. For example, the intake (Diffuser) module can be used as a stand-alone intake component, and it can be used to instantiate intake in the engine model. The intake module and its I/O are shown in Figure 2.

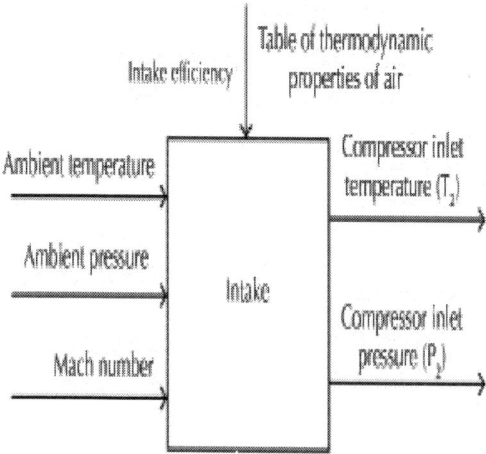

Figure 2: Intake module.

TESM includes some component modules such as: Intake, Compressor, Combustor, Turbine, and Nozzle which are modeled as lumped parameter thermodynamic systems. It means that a multiple stage compressor or turbine is simulated as one component. This approach is adopted because compressor and turbine maps are created to represent the performance of the overall component. The model can be developed to a stage-by-stage fashion if we have required authentic performance data of each component.

The modules are developed based on fundamental laws of physics such as conservation of mass, momentum, and energy. For example, the rotor dynamics is presented by the equation of conservation of angular momentum, that is based on moment of inertia of the components attached to the shaft [16].

$$\dot{N} = \frac{3600}{4\pi^2 NI} \left[W_4 C_{P4}(T_4 - T_5) - \frac{W_3 C_{P3}(T_3 - T_2)}{\eta_{mech}} \right],$$

(1)

Where N is engine spool speed, T is temperature, W is mass flow, I is engine spool moment of inertia, and n_{mech} is the engine spool mechanical efficiency.

Other most important dynamical differential equations are used in the modeling process are as follows [16]:

$$\dot{\rho} = \frac{W_3 - W_4 + u}{C_{vol}},$$

$$\dot{T}_4 = \frac{C_{P3}T_3W_3 - C_{P4}T_4W_4 + uLHV}{C_{vol}\dot{\rho}_4 T_4 C_V}.$$

(2)

In order to use individual components to create an engine model, all hardware component information such as flow, pressure ratio and efficiency maps, flow volume, and area of some components is required.

The incorporation of turbo machinery map data (compressor and turbine maps) into the MATLAB workspace (Simulink-related blocks) is a significant part of the tuning process required to match the model performance with data for a specific engine.

The other parameters to be incorporated into the MATLAB workspace are characteristic lengths, volumes, moments of inertia, design constants, efficiencies, and so forth. Values for these tuning parameters must be determined from engine design documentation, experimentation, or other simulation models. The model references these parameters at run-time.

One of TESM outputs is compressor surge margin which is highly important for safe engine operation. Compressor surge is a condition affecting both aerospace and industrial compression systems that employ turbo machinery. Surge is an unstable operating condition that can lead to the loss of an aircraft in aerospace applications and cause severe damage to industrial systems. During this compression system instability, the flow over the blades of the compressor stalls, and the pressure rise capability is reduced. When this occurs, the compressor cannot maintain the high-pressure downstream and a violent flow reversal occurs throughout the compression system.

Surge Margin (SM) of compressor refers to a margin of safety between the normal operating point of the compressor and the stability limit. Surge Margin is obtained using the incorporated turbo machinery map data, as follows [1]:

$$SM = \left(\frac{\left(P_{32}/\dot{m}_{2a_{cor}} \right)_{stall} - \left(P_{32}/\dot{m}_{2a_{cor}} \right)}{\left(P_{32}/\dot{m}_{2a_{cor}} \right)} \right), \quad P_{32} = \left(\frac{P_3}{P_2} \right).$$

(3)

The model configuration is shown in Figure 3. As can be seen, the relationships between components are easily understood and the gas path flow parameter connections between major engine components include mass flow rate, total temperature and pressure, static pressure, and Fuel-Air Ratio (FAR).

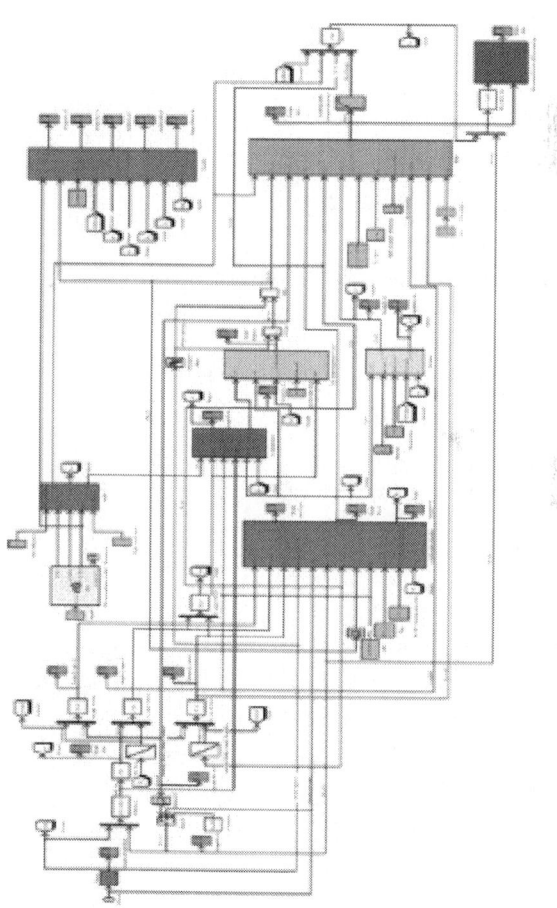

Figure 3: The Simulink model of the turbojet engine (TESM).

This model provides a basic framework for both the development of engine component modules and analysis of component interactions at the system level within the engine. Component modules can be reused and the time and cost required producing new models significantly reduced.

TESM has three main inputs including altitude and Mach number, which are ambient inputs (Flight Data), and fuel flow to the combustor chamber. The inputs are detailed below.

- Altitude: this is an ambient input with a range of 10,000 feet for this type of engine which goes into the ISA atmosphere model block.

- Mach number: this simulates the speed of aircraft and its effects on P_2, T_2, and C_a. The range for this type of engine Mach number is 0 to 0.9.

- Fuel flow: this is an important input from an engine control point of view and is set by a controller for a specific demanded Power Lever Angle (PLA).

The engine transient operation can be indicated by TESM's outputs. Table 1 presents some of these outputs.

Table 1: Turbine engine simulator model outputs

Component module	Output parameters
ISA.Atms.Model	a, , P1, T1
Intake	P2, T2
Compressor	P3, T3, W3
Combustor	P4, T4, W4
Turbine	P5, T5
Nozzle	Thrust, SFC, Ps6, Ts6, C6
Shaft	N
Surge margin model	SM

Among the model outputs, four are the most important, including: engine spool speed (N), turbine inlet temperature (T_4), compressor surge margin (SM), and engine thrust. These should be controlled by engine controllers. Figure 4 presents TESM and its more important inputs and outputs.

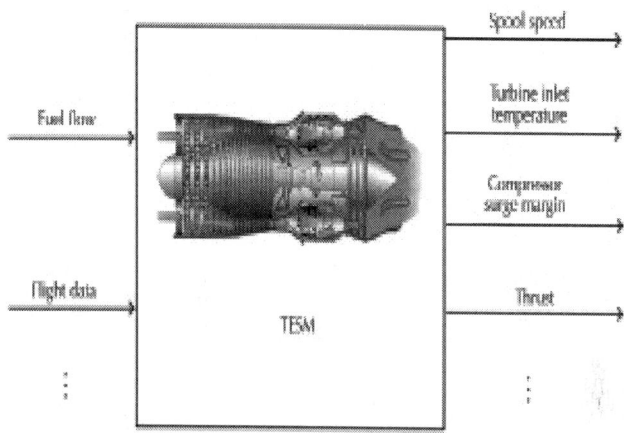

Figure 4: The turbine engine simulator model.

The model operates as a virtual test cell and enables a user to investigate "what-if" scenarios at a fraction of the cost of an engine test cell or research aircraft.

Now, to present an efficient method to estimate the fuel flow injection function to the combustor chamber which is of great importance among the engine input controlling functions (See [1, 2, 20, 21]), some different reasonable fuel flow functions are applied via the engine model, and then some important engine continuous time operation parameters (such as thrust, compressor surge margin, turbine inlet temperature, and engine spool speed, etc.) are obtained. These parameters and the fuel flow functions provide a precious database which can be used by a neural network.

Each of the foresaid fuel flow functions as an important part of the database is a series of two-dimensional vectors. Since the engine accelerates from normalized spool speed of n=0.55 to n=1.1, the starting and ending points of the functions concern the related steady state values of fuel flow. As a general rule, the Fuel-Air Ratio (FAR) put some constraints on variation of fuel flow rate. In this regard, the derivative of a fuel flow function should be restricted by some prespecified values for the engine. By considering all of the foresaid constraints and the compressor map, a series of discrete points would be obtained. These points are used to estimate the fuel flow functions.

NEURAL NETWORK TRAINING

Neural networks are a powerful technique to solve many real-world problems. They have the ability to learn from experience in order to improve their performance and to adapt themselves to changes in the environment. In addition to that, they are able to deal with incomplete information or noisy data and can be very effective especially in situations where it is not possible to define the rules or steps that lead to the solution of a problem [2].

By designing and training a feedforward multilayer perceptron neural network according to the available database; we estimate a number of different reasonable fuel flow functions providing the desired engine performance parameters such as thrust and compressor surge margin for various acceleration maneuverings. The selected engine maneuverings only include twelve fast and safe accelerations for pilot's full throttle commands to cover the critical range of engine operation. The different rates of full throttle commands will diversify these maneuverings. In fact, the proposed neural network simulates the engine performance backwardly. So a series of fuel flow functions would be obtained for various engine acceleration maneuverings. These functions will be used in the process of membership functions definition for designing a fuzzy logic controller.

The network input layer consists of four neurons which are the most important engine performance parameters and acquired by the engine model. These engine model output parameters are depicted in Figure 4. There are ten neurons in the hidden layer. The output layer includes one neuron as the fuel flow rate. The proposed neural network is shown in Figure 5.

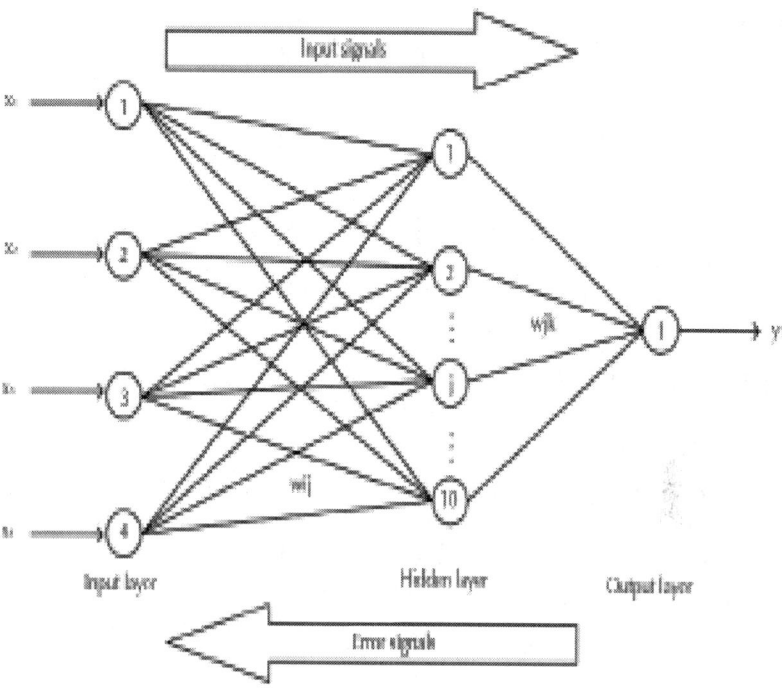

Figure 5: The three-layer back propagation neural network.

We use back propagation algorithm as a reliable and well-known algorithm to train the network. Applying this algorithm would minimize output error by changing the synapse weights.

Three hundred training data and fifty testing data are used by the network. The network activity function is as follows:

$$f(x) = \frac{1}{1 + e^{-x}}.$$

(5)

Since the above function provides normalized outputs between zero and one; so the inputs of the network should be normalized too At last, the proposed network would achieve to acceptable error by 3500 epochs and training process finish.

DESIGNING A FUEL FLOW CONTROLLER BY THE AID OF FUZZY LOGIC METHOD

Fuzzy logic is an increasingly popular method of handling systems associated with uncertainty, unmodeled dynamics, or simply where human experience is required. Its ability to deal with imprecise data can often offer an immediate benefit over conventional mathematical reasoning. It has been widely employed in control problems, particularly due to its ability to mimic the behavior of nonlinear plants. By ensuring that a properly formulated rule base is found, a fuzzy system can provide smooth transitions between operating regimes [12,22].

An FLC utilizes fuzzy logic to convert linguistic information based on expert knowledge into an automatic control strategy. In order to use the fuzzy logic for control purposes, a front-end "fuzzifier" and a rear-end "defuzzifier" are added to the usual input-output data set [11]. An FLC commonly consists of four sections: rules, fuzzifier, inference engine, and defuzzifier. Once the rule has been established, the controller can be considered as a nonlinear mapping from the input to the output. The block diagram (see Figure 6) of the generalized indistinct controller consists of four elements [23]:

- fuzzification block, transforming input physical values y_i into corresponding linguistic variables $\mu(y_i)$;
- fuzzy rule base, including:

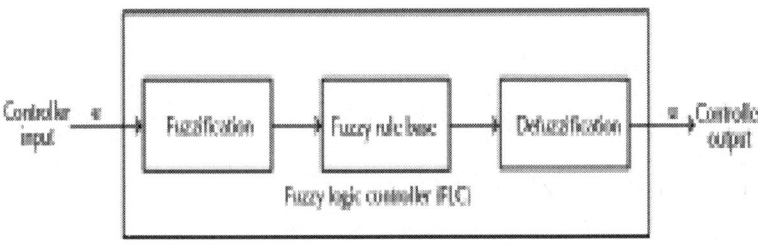

Figure 6: The block diagram of the generalized indistinct controller.

- knowledge base, containing rules table for logic output block;
- logic output block, transforming input linguistic variables into output with some belonging functions;
- defuzzification block, transforming output linguistic variables into physical control influence.

The designed FLC would determine the amount of fuel flow to the combustor chamber over its transient operation. The FLC and the available turbine engine model constitute a closed loop which is shown in Figure 7.

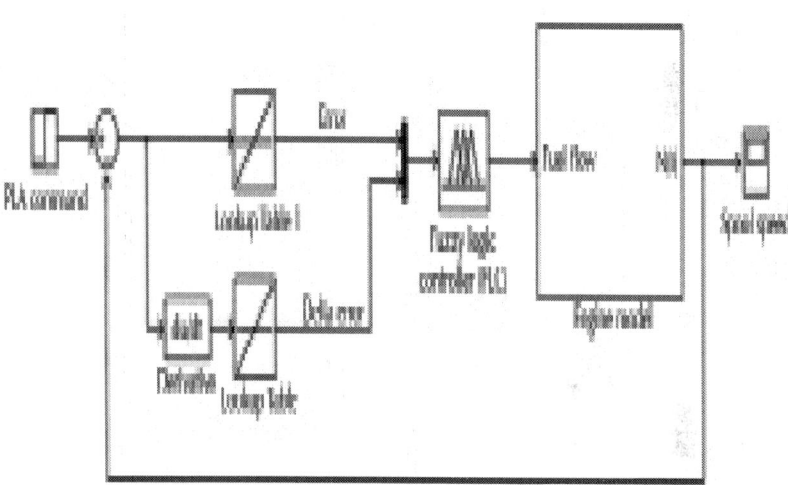

Figure 7: Layout of the turbojet engine simulator model and FLC.

The first input of controller is the spool speed error (the difference of throttle demanded spool speed and current spool speed). This input is named Error. The defined fuzzy function for the input is shown in Figure 8. The second input is the first input variations over time. This input is named Delta Error. The defined fuzzy function for the second input is shown in Figure 9.

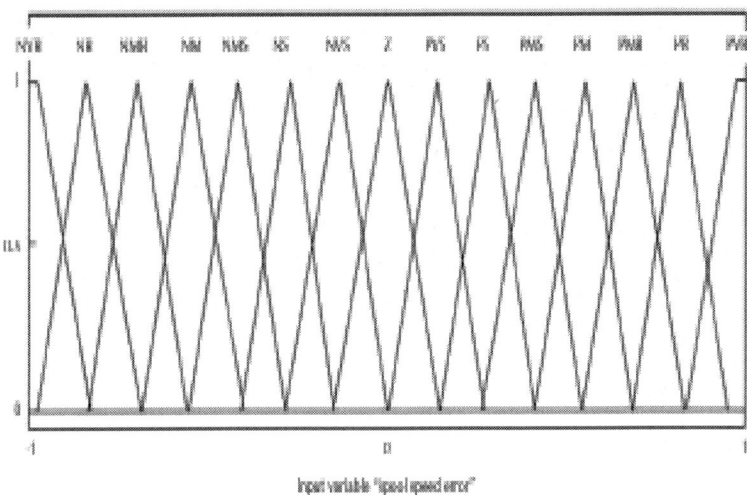

Figure 8: Error membership functions.

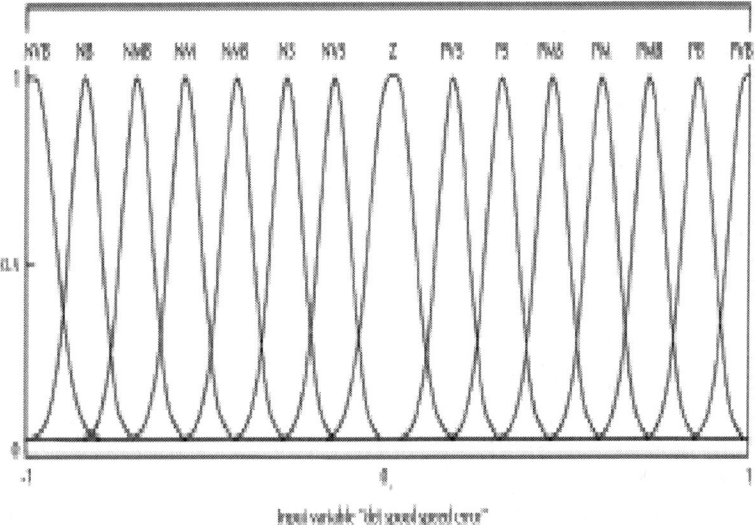

Figure 9: Delta error membership functions.

Defining the proper fuzzy membership functions for fuel flow has a great influence on the controller behavior. So, the obtained desired fuel flow functions as the neural network results would be helpful for defining the fuel flow fuzzy membership functions. This approach is helpful to defuzzify the fuzzy fuel flow membership functions.

In this regard, each of the obtained desired fuel flow functions is separately considered over fifteen equispaced time intervals in [0.0, 3.5]. So an equal number of nonequispaced intervals are also obtained on each fuel flow axis. Since the rate of fuel flow injection is much higher over the starting time intervals (starting phase of acceleration), so a large number fuel flow intervals are situated close to the end part of their axis. Fuel flow intervals on a sample desired fuel flow function are depicted by different arrows in Figure 10.

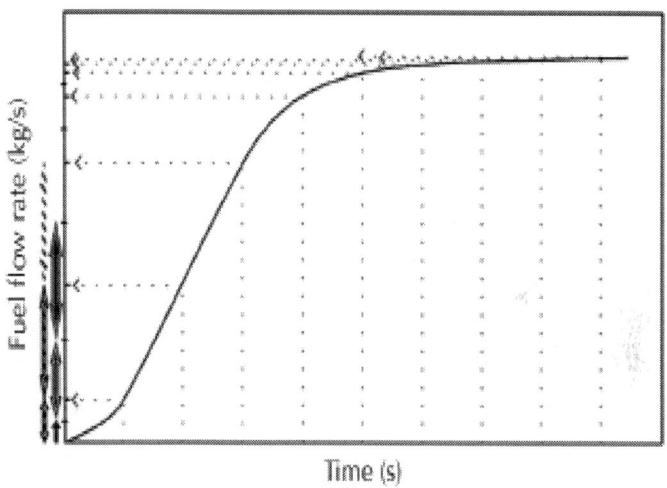

Figure 10: Fuel flow interval construction on a desired sample fuel flow function.

Using a simple mapping method, each of these intervals creates a triangular fuel flow membership function. According to the proposed mapping, the membership degree for midpoint of an interval is one and for endpoints is zero. With respect to the proposed arrangement of intervals in Figure 10, the first three of fuel flow fuzzy membership functions are presented in Table 2.

Table 2: The first three of fuel flow intervals and the related membership degrees

Function name	NVB			NB			NMB		
Fuel interval points	First	Central	End	First	Central	End	First	Central	End
Fuel amounts	—	0.036	0.041	0.036	0.042	0.047	0.042	0.051	0.062
Membership degree	—	1.0	0.0	0.0	1.0	0.0	0.0	1.0	0.0

A similar approach is applied for each desired fuel flow function. Then, the obtained fuzzy fuel flow functions are used to define a final set of fuel flow fuzzy membership functions. In this regard, for each fuel interval the Arithmetic mean of fuel values with the highest membership degree is obtained and labeled with maximum membership degree. These final membership functions are used in designing procedure of the FLC.

Conventionally, fuzzy rules are established by a combination of knowledge, experience, and observation and may thus not be optimal. Additionally, in spite of efforts to formalize a development standard for fuzzy controllers, fine tuning its performance is still a matter of trial and error [5, 12, 22].

A variety of membership functions can be defined for inputs and output of the controller. By changing the number and types of the functions and rules, the engine behavior would change.

The acceleration time to achieve pilot demanded operation point is of great importance. This time should be short as much as possible [16]. On the other hand, if the acceleration time be shorter than its allowable limit, compressor surge would be likely. So, it is necessary for the FLC to limit the inordinate engine acceleration.

In this paper, the applied logic for creation the FLC rules includes higher spool acceleration rates during lower spool speeds and lower spool acceleration rates during higher spool speeds. Some selected rules which produce the central part of control surface are presented in Table 3.

Table 3: Selected fuzzy rules

Error	Delta-Error						
	NMS	NS	NVS	Z	PVS	PS	PMS
NMS	NMS	NS	NS	NS	NS	NVS	Z
NS	NMS	NS	NS	NS	NVS	Z	PVS
NVS	NMS	NS	NS	NVS	Z	PVS	PS
Z	NMS	NS	NVS	Z	PVS	PS	PMS
PVS	NMS	NS	NVS	Z	PVS	PS	PMS
PS	NVS	Z	Z	Z	PVS	PS	PMS
PMS	Z	PVS	PS	PS	PS	PS	PMS

The defuzzification process, which takes place after the generation of the fuzzy control signals, is completed using the inference mechanism. The resulting fuzzy set must be converted to a quantity which would be sent to the process regulating valve as a control signal. In this part, the inference results of all activated logic rules are synthesized into crisp output for making a decision. In this study, the logic AND has been implemented with the minimum operator, and the defuzzification method is based on bisector area.

The variation of fuel flow versus the two controller inputs is depicted in Figure 11.

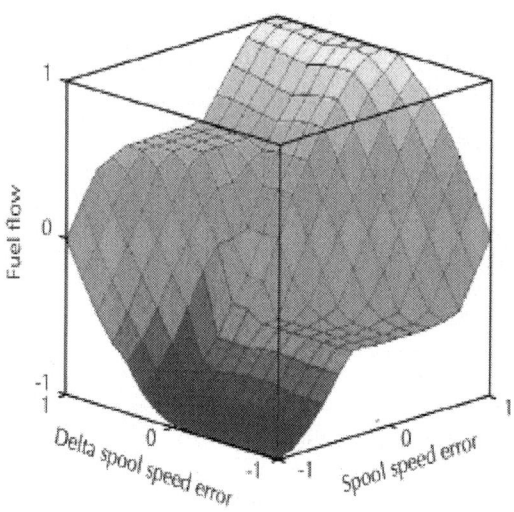

Figure 11: Control surface output of the Fuzzy Logic Controller (FLC).

By designing the controller, the influence of controller parameters on turbine engine should be examined in more detail. So it would be essential to simulate the turbine engine and controller simultaneously.

TESTING AND DEMONSTRATION

The turbine engine acceleration for full throttle command would be considered to achieve the most critical condition. The simulations are run with the unit step power lever angle and constant zero values for

Mach number and altitude.

The desirable engine performance would be obtained by applying more and various types of membership functions and rules and then rerunning the simulation program. This technique is effective to achieve the best fuzzy logic controller. The simulation results of engine model with FLC are presented in Figure 12.

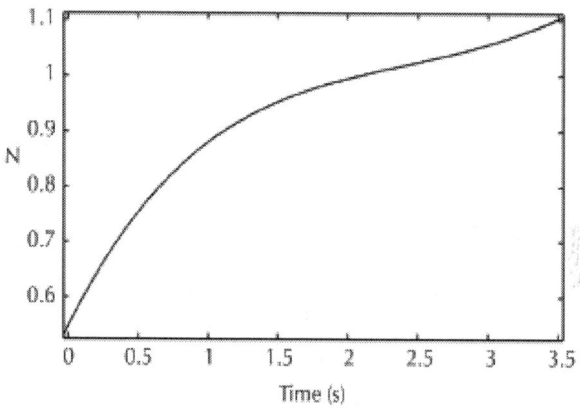

(a)

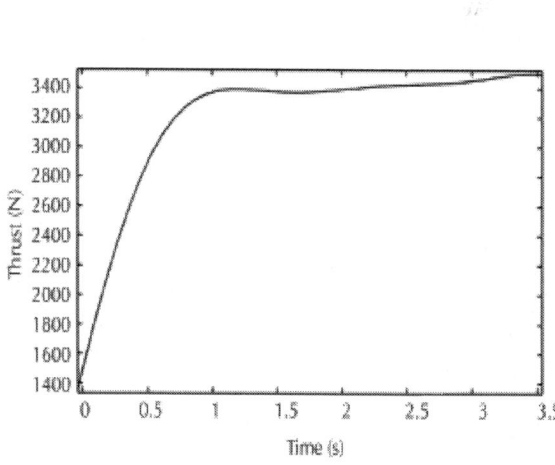

(b)

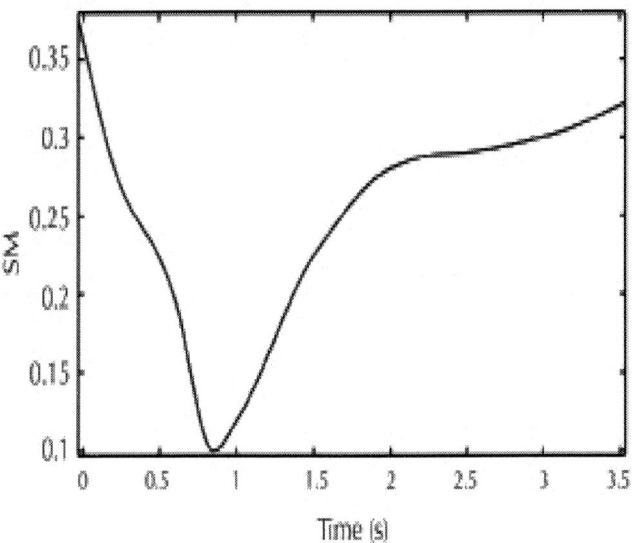

(c)

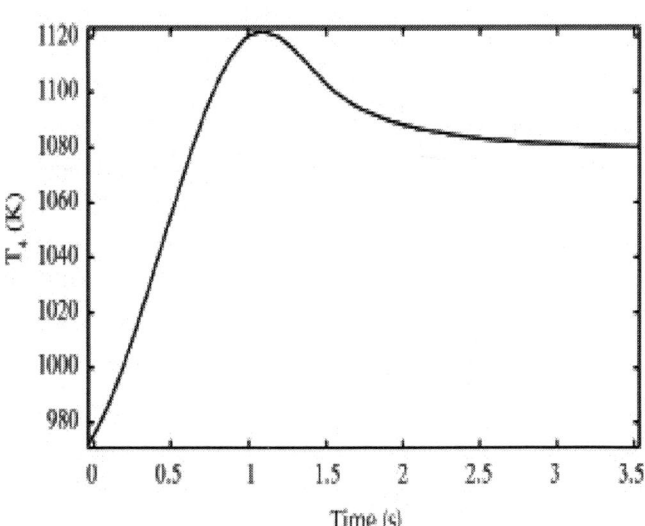

(d)

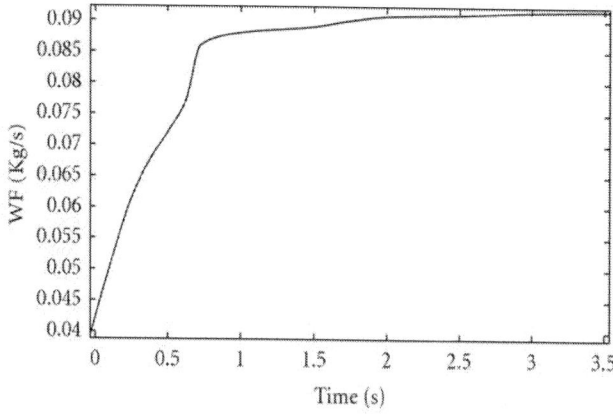

(d)

Figure 12: Trace of engine model responses with PLA Movement at sea level, M=0 versus simulation time (s). (a) Normalized rotational speed. (b) Thrust (N). (c) Surge margin. (d) Turbine inlet temperature (K). (e) Fuel flow.

When the turbine output power overcomes the compressor input power, the acceleration operation will be done. So, increasing the fuel flow rate is a common method to accelerate the engine spool.

Through the initial time of acceleration, the turbine inlet temperature and the compressor output pressure will be risen and the compressor air flow rate reduced. Considering the said circumstances during engine acceleration, high acceleration rate may lead to compressor surge or stall and excessive turbine inlet temperature which must be controlled by FLC. Figures 12(c) and 12(d) demonstrate the FLC capability in providing the above requirements.

The acceptability of gas turbine engine is crucially influenced by its ability in producing a demanded thrust while desirable fuel consumption. The demanded thrust for the jet engine is 3500 (N). Figure 12(b) shows the FLC capability in reaching the demanded thrust.

The simulation results of engine model with fuzzy controller in comparison with the engine model with a conventional controller are presented in Table 4.

Table 4: Simulation results of engine model with FLC in comparison with the engine model with conventional controller

Parameter	Fuzzy logic controller (simulation)					Conventional controller (simulation)				
Time (Sec)	0	0.8	1.6	2.4	3.2	0	0.8	1.6	2.4	3.2
Spool speed (normalized)	0.55	0.86	0.96	1.02	1.05	0.55	0.74	0.84	0.95	0.98
Turbine inlet temperature (K)	975.39	1118	1098	1086	1081	975.39	1129	1119	1109	1103
Thrust (N)	1482	3404	3462	3471	3478	1482	3379	3393	3409	3412
SM	0.36	0.1	0.23	0.29	0.31	0.36	0.12	0.26	0.33	0.35

As it is seen from Table 4, the proposed FLC has led to more rapid acceleration and lower turbine inlet temperature. However, the SM is reduced in comparison with the conventional controller.

One of important inputs for FLC is the engine spool speed error which is shown in Figure 13. The magnitude of this error varies from its maximum (0.55) at the start of simulation to its minimum (zero) by achieving the demanded spool speed.

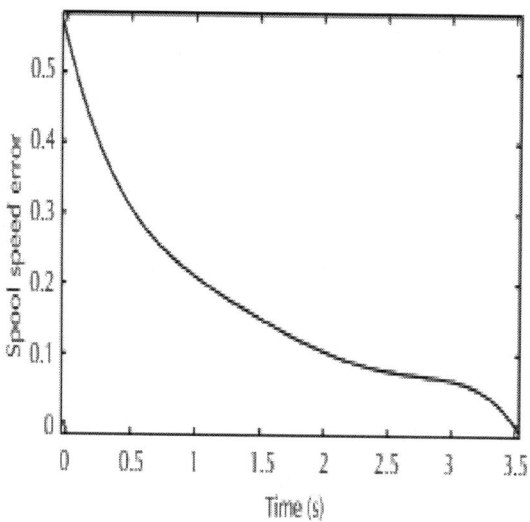

Figure 13: Trace of engine spool speed error.

At the end of specified engine acceleration, some parameters such as time of engine acceleration and thrust are presented in Table 5. The acceleration time has reduced by FLC. This would be an important advantage of the proposed FLC.

Table 5: Simulation results of engine model with fuzzy controller in comparison with the engine testing operation

Parameter	Fuzzy logic controller (simulation)	Conventional controller (test)
	N=1.1	N=1.1
Time (Sec)	3.5	4.2
Thrust (N)	3484	3420

Whether a fuzzy control design will be stable is a somewhat open question. The problems of FLC stability analysis and optimality are not addressed explicitly; such issues are still open problems in fuzzy controller design. Various nonlinear stability analysis methodologies could be applied for analyzing fuzzy control systems. Some methods have been proposed based on the Lyapunov's second method. The second method of Lyapunov (also referred to as the direct method) is the most general for determining the stability of a nonlinear and/ or time-variant system of any order. It is a hot research topic now. In this project, the response curves of the system and its parameters including rise time, overshoot ratio, and settle time may be analyzed for specifying the performance of the FLC. The proper parameters for system response are obtained by adjusting either the rules, the input and output scaling factors, or some other parameters of the FLC. Using the foresaid method, the control loop behavior for the system presented in this paper is well known and considered sufficient.

CONCLUSIONS

The rationale behind this study was the need to develop a new technique that may be able to improve the performance, and simultaneously enhance the flexibility, of the control strategy for future concepts in aero-engines. These engines include nonconventional engines or multiregime engines which are used by modern military aircraft with more complex maneuvers.

As engines become more complex and have more controllable and measurable parameters, the need for such technique will increase if they are to achieve their full potential.

The steady and transient simulations indicated the fuel flow importance as a comprehensive controlling parameter which has significant effects on all engine performance parameters. So the control system featured loops to prevent engine over speeds, compressor surge, and check turbine inlet temperature limit by scheduling the fuel flow during accelerations and decelerations. The spool speed error and its differential over time were used as fuzzy controller inputs. The influence of controller parameters on the turbine engine performance was analyzed by changing the number and types of the membership functions and rules. This is effective to design the best controller. The designed neural network simulated the engine performance backwardly. So it empowered us to obtain a series of fuel flow functions in various engine acceleration maneuverings. These functions were used in the process of membership functions definition.

The results obtained from full throttle simulation of the turbine engine and controller proved logical and well founded. The compressor surge margin is safe, and turbine inlet temperature is restricted by lower allowable value.

The simulation results of engine model with fuzzy controller illustrate that the proposed controller achieves the desired performance.

REFERENCES

1. M. Bazazzade, H. Badihi, and A. Shahriari, "Improved turbine engine hierarchical modeling and simulation based on engine fuel control system," Journal of Aerospace Science & Technology, vol. 6, pp. 45–53, 2009.

2. H. Badihi, A. Shahriari, and A. Naghsh, "Artificial neural network application to fuel flow function for demanded jet engine performance," in Proceedings of the IEEE Aerospace Conference, IEEE, Big Sky, Mont, USA, March 2009. ·

3. L. C. Jaw and S. Garg, "Propulsion Control Technology Development in the United States (A Historical Perspective)," NASA, Glenn Research Center, Cleveland, Ohio, USA, 2005.

4. J. S. Litt, D. L. Simon, S. Garg, et al., "A Survey of Intelligent Control and Health Management Technologies for Aircraft Propulsion Systems," NASA, Glenn Research Center, Cleveland, Ohio, USA, 2005.

5. P. J. Antsaklis and K. M. Passino, An Introduction to Intelligent and Autonomous Control, Kluwer Academic, Boston, Mass, USA, 1993.

6. L. A. Zadeh, "A fuzzy-set-theoretical interpretation of linguistic hedges," Journal of Cybernetics, vol. 2, no. 3, pp. 4–34, 1972.

7. M. M. Gupta, Advances in Fuzzy Set Theory and Application, North-Holland, Amsterdam, The Netherlands, 1987.

8. B. Kosko, Neural Networks and Fuzzy Systems: A Dynamical Systems Approach to Machine Intelligence, Prentice Hall, New York, NY, USA, 1991.

9. L. A. Zadeh, G. J. Klir, and B. Yuan, Fuzzy Sets, Fuzzy Logic, Fuzzy Systems: Selected Papers, World Scientific Publishing, River Edge, NJ, USA, 1996.

10. L. A. Zadeh, "Fuzzy sets," Information and Control, vol. 8, no. 3, pp. 338–353, 1965.

11. E. H. Mamdani and S. Assilian, "An experiment in linguistic synthesis with a fuzzy logic controller,"International Journal of Man-Machine Studies, vol. 7, no. 1, pp. 1–13, 1975.

12. A. J. Chipperfield, B. Bica, and P. J. Fleming, "Fuzzy scheduling control of a gas turbine aero-engine: a multiobjective approach," IEEE Transactions on Industrial Electronics, vol. 49, no. 3, pp. 536–548, 2002. · ·

13. A. Zilouchian, M. Juliano, T. Healy, and J. Davis, "Design of a fuzzy logic controller for a jet engine fuel system," Control Engineering Practice, vol. 8, no. 8, pp. 873–883, 2000. · ·

14. J. W. Kim and S. W. Kim, "Design of incremental fuzzy PI controllers for a gas-turbine plant,"IEEE/ASME Transactions on Mechatronics, vol. 8, no. 3, pp. 410–414, 2003. · ·

15. A. Martucci and A. J. Volponi, "Fuzzy fuel flow selection logic for a real time embedded full authority digital engine control," Journal of Engineering for Gas Turbines and Power, vol. 125, no. 4, pp. 909–916, 2003. · ·

16. P. P. Walsh and P. Fletcher, Gas Turbine Performance, Blackwell, Malden, Mass, USA, 2004.

17. A. Behbahani, R. K. Yedavalli, P. Shankar, and M. Siddiqi, "Modeling, diagnostics and prognostics of A two-spool turbofan engine," in Proceedings of the 41st AIAA/ASME/SAE/ASEE Joint Propulsion Conference and Exhibit (AIAA ‹05), Tucson, Ariz, USA, July 2005.

18. G. Crosa, F. Pittaluga, A. Trucco, F. Beltrami, A. Torelli, and F. Traverse, "Heavy-duty gas turbine plant aerothermodynamic simulation using simulink," Journal of Engineering for Gas Turbines and Power, vol. 120, no. 3, pp. 550–555, 1998.

19. M. Lichtsinder and Y. Levy, "Jet engine model for control and real-time simulations," Journal of Engineering for Gas Turbines and Power, vol. 128, no. 4, pp. 745–753, 2006. · ·

20. K. Lietzau and A. Kreiner, "Model Based Control Concepts for Jet Engines," ASME Paper, 2001.

21. K. Lietzau and A. Kreiner, "The Use of Onboard Real-Time Models for Jet Engine Control," MTU Aero Engine Germany, 2004.

22. G. M. Nelson and H. Lakany, "An investigation into the application of fuzzy logic control to industrial gas turbines,"

Journal of Engineering for Gas Turbines and Power, vol. 129, no. 4, pp. 1138–1142, 2007. · ·

23. L. Reznik, Fuzzy Controllers, NEWNES, Melbourne, Australia, 1997.

Citations

CHAPTER 1

I. Gurrappa, A. Gogia and I. Yashwanth, "The Behaviour of Superalloys in Marine Gas Turbine Engine Conditions," Journal of Surface Engineered Materials and Advanced Technology, Vol. 1 No. 3, 2011, pp. 144-149. doi: 10.4236/jsemat.2011.13022.

CHAPTER 2

Alfonso Calabria, Roberto Capata, Mario Di Veroli, and Gianluca Pepe, Testing of the Ultra-Micro Gas Turbine Devices (1 - 10 kW) for Portable Power Generation at University of Roma 1: First Tests Results, doi:10.4236/eng.2013.55058.

CHAPTER 3

Zhi-tao Wang, Ning-bo Zhao, Wei-ying Wang, Rui Tang, and Shu-ying Li, "A Fault Diagnosis Approach for Gas Turbine Exhaust Gas Temperature Based on Fuzzy C-Means Clustering and Support Vector Machine,"Mathematical Problems in Engineering, Article ID 240267, in press.

CHAPTER 4

R. K. Giridhar, P. V. Ramaiah, G. Krishnaiah, and S. G. Barad, "Gas Turbine Blade Damper Optimization Methodology," Advances in Acoustics and Vibration, vol. 2012, Article ID 316761, 13 pages, 2012. doi:10.1155/2012/316761.

CHAPTER 5

Weiying Wang, Zhiqiang Xu, Rui Tang, Shuying Li, and Wei Wu, "Fault Detection and Diagnosis for Gas Turbines Based on a Kernel-ized Information Entropy Model," The Scientific World Journal, vol. 2014, Article ID 617162, 13 pages, 2014, doi:10.1155/2014/617162.

CHAPTER 6

Ufot, E. , Douglas, I. and Hart, H. (2014) Surface Temperatures Determination with Influencing Convective and Radiative Thermal Resistance Parameters of Combustor of Gas Turbine. Engineering, 6, 550-558. doi:10.4236/eng.2014.69056.

CHAPTER 7

Ning-bo Zhao, Xue-you Wen, and Shu-ying Li, "Dynamic Time-Delay Characteristics and Structural Optimization Design of Marine Gas Turbine Intercooler," Mathematical Problems in Engineering, vol. 2014, Article ID 701843, 14 pages, 2014. doi:10.1155/2014/701843.

CHAPTER 8

M. Bazazzadeh, H. Badihi, and A. Shahriari, "Gas Turbine Engine Control Design Using Fuzzy Logic and Neural Networks," International Journal of Aerospace Engineering, vol. 2011, Article ID 156796, 12 pages, 2011. doi:10.1155/2011/156796.

Index

A

Artificial neural networks (ANN)
43

B

Backpropagation (BP) 63
Blue colour 32

C

Coulomb's law 78

D

Data acquisition and local monitoring subsystem (DALM)
114

Data communication subsystem
(DAC) 114
Data management subsystem
(DMS) 114
Ddirected acyclic graph (DAG)
53

E

Electrical power 28, 29
Electrochemical model 14
Electron Dispersive Spectroscopy
(EDS) 5
Electronics (ECU) 20
Exhaust gas temperature (EGT)
41, 42

F

Friction damping concept 69, 70
Fuel-Air Ratio (FAR) 203, 205
Fuel-Air Ratio (FAR). 203
Fuzzy C-means (FCM) 41
Fuzzy Logic Controller (FLC)
    195, 197, 214
Fuzzy logic (FL), 197

G

Gas turbine 109, 110, 113, 196
General regression neural net-
    work (GRNN) 43
genetic algorithm (GA) 181
Genetic algorithms (GA), 197
Good result 137

I

Inlet Guide Vane (IGV) 198
Intelligent diagnosis system (IDS)
    114
Intercooled regenerated (ICR)
    156

M

Micro-scale electric 18
Model. Generalization 134

N

Neural networks (NN), 197

P

Particle swarm optimization
    (PSO) 181
Power generation 109, 111
Power Lever Angle (PLA). 204
Probabilistic reasoning (PR). 197

R

Radial basis function (RBF) 53
Recent extensive research 15

S

Scanning Electron Microscope
    (SEM) 5
Simulated annealing (SA) 158,
    186, 191
Support vector machine (SVM)
    41, 51
Surface temperature 141
Surge margin (SM), 204

T

Thermal conductivity 176, 179,
    191
TIT (Turbine inlet temperature)
    19
Turbine inlet temperature (TIT)
    25